# Dictionary of Food Ingredients

## Fourth Edition

**Robert S. Igoe, MS, MBA**
Director (Retired), Latin America
Kelco Alginates
A Division of Monsanto Company
San Diego, California

**Y. H. Hui, PhD**
President
Science and Technology System
West Sacramento, California

AN ASPEN PUBLICATION®
Aspen Publishers, Inc.
Gaithersburg, Maryland
2001

The authors have made every effort to ensure the accuracy of the information herein. However, appropriate information sources should be consulted, especially for new or unfamiliar procedures. It is the responsibility of every practitioner to evaluate the appropriateness of a particular opinion in the context of actual clinical situations and with the due considerations to new developments. The author, editors, and the publisher cannot be held responsible for any typographical or other errors found in this book.

Aspen Publishers, Inc., is not affiliated with the American Society of Parenteral and Enteral Nutrition.

Library of Congress Cataloging-in-Publication Data

Igoe, Robert S.
Dictionary of food ingredients / Robert S. Igoe, Y.H. Hui—4th ed.
p. cm.
Includes bibliographical references.
ISBN 0-8342-1952-2 (pbk.)
1. Food—Composition—Dictionaries. I. Hui, Y.H. (Yiu H.) II. Title.
TX551.I26 2001
641'.03—dc21
00-053578

Orders: (800) 638-8437
Customer Service: (800) 234-1660

**About Aspen Publishers** • For more than 40 years, Aspen has been a leading professional publisher in a variety of disciplines. Aspen's vast information resources are available in both print and electronic formats. We are committed to providing the highest quality information available in the most appropriate format for our customers. Visit Aspen's Internet site for more information resources, directories, articles, and a searchable version of Aspen's full catalog, including the most recent publications: **www.aspenpublishers.com**
**Aspen Publishers, Inc.** • The hallmark of quality in publishing
Member of the worldwide Wolters Kluwer group.

Editorial Services: Erin McKindley
Library of Congress Catalog Card Number: 00-053578
ISBN: 0-8342-1952-2

*Printed in the United States of America*

1 2 3 4 5

# Table of Contents

# Preface

The *Dictionary of Food Ingredients* is a unique, easy-to-use source of information on over 1,000 food ingredients. Like the previous editions, the new and updated Fourth Edition provides clear and concise information on currently used additives, including natural ingredients, FDA-approved artificial ingredients, and compounds used in food processing. The dictionary entries, organized in alphabetical order, include information on ingredient functions, chemical properties, and uses in food products. The updated and revised Fourth Edition contains new entries, an ingredient categories section grouping ingredients by type and comparative properties, a listing of food ingredients under the U.S. Code of Federal Regulations, a listing of respective European E numbers, and a bibliography section.

Users of the three previous editions have commented favorably on the dictionary's straightforward and clearly written definitions, and we have endeavored to maintain that standard in this new edition. We trust it will continue to be a valuable reference for the food scientist, food processor, food product developer, nutritionist, extension specialist, and student.

# PART I

# Ingredients Dictionary

# A

**Acacia**—*See Arabic.*

**Acesulfame-K**—A non-nutritive sweetener also termed acesulfame potassium. It is a white, crystalline product that is 200 times sweeter than sucrose. It is not metabolized in the body. It has some metallic off-tastes. It is readily soluble and heat and acid stable. It provides a synergistic sweetening effect combined with other sweeteners. It is used in beverages, desserts, confectionery, and bakery products.

**Acesulfame Potassium**—*See Acesulfame-K.*

**Acetanisole**—A solid, pale yellow flavoring agent with a Hawthorn-like odor. It is soluble in most fixed oils and propylene glycol, and it is insoluble in glycerine and mineral oil. It is obtained by chemical synthesis. This flavoring substance or its adjuvant may be safely used in food in the minimum quantity required to produce its intended flavor. It can be used alone or in combination with other flavoring substances or adjuvants. It is also termed p-methoxyacetophenone.

**Acetic Acid**—An acid produced chemically from the conversion of alcohol to acetaldehyde to acetic acid. It is the principal component of vinegar which contains not less than 4 g of acetic acid in 100 cm$^3$ at 20°C. The approved salts include sodium acetate, calcium acetate, sodium diacetate, and calcium diacetate. It is used as a preservative, acidulant, and flavoring agent in catsup, mayonnaise, and pickles. It can be used in conjunction with leavening agents to release carbon dioxide from sodium bicarbonate.

**Acetic Acid, Glacial**—*See Glacial Acetic Acid.*

**Acetic Anhydride**—An esterifier for food starch; also used in combination with adipic anhydride.

**Acetone Peroxide**—A dough conditioner, maturing, and bleaching agent that is a mixture of monomeric and linear dimeric acetone peroxides which are strong oxidizing agents. It is used to age and bleach flour.

**Acetylated Monoglyceride**—An emulsifier manufactured by the interesterification of edible fats with triacetin in the presence of catalysts or by the direct acetylation of edible monoglycerides with acetic anhydride without the use of catalysts. It is characterized by sharp melting points, stability to oxidative rancidity, film forming, stabilizing, and lubricating properties. It is used as a protective coating for meat products, nuts, and fruits to improve their appearance, texture, and shelf life. The coatings are applied by spraying, panning, and dipping. It is used in cake shortening and fats for whipped topping to enhance the aeration and improve foam stabilization. It is found in dry-mix whipped topping.

**Acetylated Tartaric Acid Monoglyceride**—*See Diacetyl Tartaric Acid Esters of Mono- and Diglycerides.*

**Acetyl Tartrate Mono- and Diglyceride**—*See Diacetyl Tartaric Acid Esters of Mono- and Diglycerides.*

**Acid Calcium Phosphate**—*See Monocalcium Phosphate.*

**Acid Casein**—The principle milk protein which is prepared from skim milk by precipitation with an acid, such as lactic, sulfuric, or hydrochloric acid, to lower the pH of the milk to 4.4 to 4.7. Caseins are identified according to the type of acid used, but in this form have little utility in foods, though they are used to some extent in cereal and bread fortification. Neutralization of the caseins yields the salts of which sodium and calcium caseinates are the most common. *See Casein.*

**Acid-Modified Corn Starch**—*See Cornstarch, Acid-Modified.*

**Acid Sodium Pyrophosphate**—*See Sodium Acid Pyrophosphate.*

**Acidulants**—Acids used in processed foods for a variety of functions that enhance the food. They are used as flavoring agents, preservatives in microbial control, chelating agents, buffers, gelling and coagulating agents, and in many other ways.

**Aconitic Acid**—A flavoring substance which occurs in the leaves and tubers of *Aconitum napellus* L. and other *Ranunculaceae.* Transaconitic acid can be isolated during sugar cane processing, by precipitation as the calcium salt from cane sugar or molasses. It may be synthesized by sulfuric acid dehydration of citric acid but not by the methanesulfonic acid method. It is used in a maximum level, as served, of 0.003 percent for baked goods, 0.002 percent for alcoholic beverages,

0.0015 percent for frozen dairy products, 0.0035 percent for soft candy, and 0.0005 percent or less for all other food categories.

**Acrolein**—This is used in the ether etherification of food starch up to 0.6 percent and for the esterification and etherification of food starch up to 0.3 percent with vinyl acetate up to 7.5 percent.

**Adipic Acid**—An acidulant and flavoring agent. It is characterized as stable, nonhygroscopic, and slightly soluble, with a water solubility of 1.9 g per 100 ml at 20°C. It has a pH of 2.86 at 0.6 percent usage level at 25°C. It is used in powdered drinks, beverages, gelatin desserts, lozenges, and canned vegetables. It is also used as a leavening acidulant in baking powder. It can be used as a buffering agent to maintain acidities within a range of pH 2.5 to 3.0. It is occasionally used in edible oils to prevent rancidity.

**Adipic Anhydride**—An esterifier for food starch in combination with acetic anhydride.

**Agar**—A gum obtained from red seaweeds of the genera *Gelidium*, *Gracilaria*, and *Eucheuma*, class Rhodophyceae. It is a mixture of the polysaccharides agarose and agaropectin. It is insoluble in cold water, slowly soluble in hot water, and soluble in boiling water, forming a gel upon cooling. The gels are characterized as being tough and brittle, setting at 32 to 40°C and melting at 95°C. A rigid, tough gel can be formed at 0.5 percent. Agar mainly functions in gel formation because of its range between melting and setting temperatures, being used in piping gels, glazes, icings, dental impression material, and microbiological plating. Typical use levels are 0.1 to 2.0 percent.

**Agar-Agar**—*See Agar*.

**Agave Nectar**—A sweetener obtained from the core of the Blue Agave (botanical name: *Agave tequilana* Weber). It is predominantly fructose and is approximately 30 percent sweeter than sugar on a relative sweetness basis. It is a good source of inulin. It functions as a sweetener, flavor enhancer, and fermentation aid.

**Albumin**—Any of several water-soluble proteins that are coagulated by heat and are found in egg white, blood serum, and milk. Milk albumin is termed lactalbumin and milk albuminate and it contains 28 to 35 percent protein and 38 to 52 percent lactose. It is used as a binder in imitation sausage, soups, and stews.

**Aldehyde C-9**—*See Nonanal*.

**Aldehyde C-16—*See Ethyl-Methyl-Phenyl-Glycidate.***

**Aldehyde C-18—*See (Gamma)-Nonalactone.***

**Algin**—Gum derived from alginic acid which is obtained from brown seaweed genera, such as *Macrocystis pyrifera*. The derivatives are sodium, ammonium, and potassium alginates of which the sodium salt is most common. They are used to provide thickening, gelling, and binding. A derivative designed for improved acid and calcium stability is propylene glycol alginate. The algins are soluble in cold water and form non-thermoreversible gels in reaction with calcium ions and under acidic conditions. Algin is used in ice cream, icings, puddings, dessert gels, and fabricated fruit.

**Alginate**—A gum derived from alginic acid that is used to provide thickening, gelling, and binding. ***See Algin.***

**Alginic Acid**—The acidic, insoluble form of algin that is a white to yellowish fibrous powder obtained from brown seaweed genera, such as *Macrocystis pyrifera*. The derivatives are soluble and include sodium, potassium, and ammonium alginate and propylene glycol alginate. It is used as a tablet disintegrant and as an antacid ingredient.

**All-Purpose Flour**—A flour that is intermediate between long-patent flours (bread flour) which contain more than 10.5 percent protein and 0.40 to 0.50 percent ash and short-patent flours (cake flour) which generally contain less than 10 percent protein and less than 0.40 percent ash. It is made from hard or soft wheat and is used in baking and in gravies. It is also termed family flour.

**Allspice**—A spice made from the dried, nearly ripe berries of *Pimenta officinalis*, a tropical evergreen tree. It has an aroma and flavor resembling that of a blend of nutmeg, cinnamon, and cloves. For labeling purposes, allspice refers to the spice of Jamaican origin. It is used in fruit pies, cakes, mincemeat, plum pudding, soups, and sauces.

**Allyl Anthranilate**—A synthetic flavoring agent that is a light yellow colored liquid of green leaf-wine odor. It is stable but may cause discoloration. It should be stored in glass or tin containers. It is used as flavoring for its wine note and has application in beverages and candy at 1 to 2 parts per million.

**Allyl Caproate—*See Allyl Hexanoate.***

**Allyl Cinnamate**—A synthetic flavoring agent that is a fairly stable, hazy, colorless to light yellow colored liquid of cherry odor. It is used for its cherry note in flavors and has application in baked goods and candies at 1 to 2 parts per million.

**Allyl-2,4-Hexadienoate**—*See Allyl Sorbate.*

**Ally Hexanoate**—A liquid flavoring agent with a strong pineapple odor and pale yellow color. It is practically insoluble in propylene glycol and miscible with alcohol, most fixed oils, and mineral oil. It is obtained by chemical synthesis. It can be used alone or in combination with other flavoring substances or adjuvants. It is also termed allyl caproate.

**Allyl Isothiocyanate**—A synthetic flavoring agent that is a moderately stable, colorless to pale yellow liquid of pungent and irritating odor. It should be stored in glass containers. It is used as an artificial oil of mustard and as an imitation horseradish flavor with application in condiments, meats, and pickles at 87 parts per million. It is also termed mustard oil.

**Allyl Isovalerate**—A synthetic flavoring agent that is a stable, colorless to light yellow liquid of fruity odor. It should be stored in glass or tin containers. It has usage in fruit flavors with applications in beverages, baked goods, ice cream, and candy at 8 to 50 parts per million.

**Allyl Mercaptan**—A synthetic flavoring agent that is a stable, colorless liquid of garlic-like odor. It should be stored in glass or tin containers. It is used in artificial garlic flavors for application in condiments at 3 parts per million, and in baked goods at 2 parts per million. It is also termed 2 propylene-1 thiol.

**Allyl Nonanoate**—A synthetic flavoring agent that is a stable, colorless to light yellow liquid of fruity-cognac odor. It should be stored in glass or tin containers. It is used in fruit flavors like melon and pineapple for application in candy, ice cream, and beverages at 0.70 to 5 parts per million.

**Allyl Octanoate**—A synthetic flavoring agent that is a colorless to light yellow liquid and has a fruity odor. It is alkali and mineral acid unstable and should be stored in glass, tin, or resin-lined containers. It is used to give flavors a fruity note and has application in dessert gels, puddings, beverages, and candy at 3 to 25 parts per million.

**Allyl Phenoxyacetate**—A synthetic flavoring agent that is a stable, colorless to light yellow liquid of heavy fruit note odor. It should be stored in glass or tin containers. It is used in pineapple, quince, and fruit flavors with applications in candy and beverages at 1 to 3 parts per million.

**Allyl Phenylacetate**—A synthetic flavoring agent that is a stable, colorless to light yellow liquid with a fruity odor of banana and honey. It should be stored in glass or tin containers. It is used in flavors for honey and has application in candy and baked goods at 10 to 15 parts per million.

**Allyl Sorbate**—A synthetic flavoring agent that is a colorless to light yellow liquid of sharp fruity odor. It is subject to polymerization and should be stored in glass or tin containers. It is used in pineapple and other fruit flavors which have application in puddings, candy, and beverages at 1 to 2 parts per million. It is also termed allyl-2,4-hexadienoate.

**Almond**—A nut obtained from the almond tree *Prunus amygdalus*. It exists as a sweet or bitter variety, with the sweet variety being used as edible nuts and the bitter variety being used as a source of almond oil. The obtainable forms range from whole nuts to slices to powder. The nuts are used as snacks, as a garnish on pastries, and as a flavorant.

**Almond Oil**—The oil of the bitter almond after the removal of hydrocyanic acid. It is a colorless to slightly yellow liquid having a strong principally almond-like aroma. It is used mainly in the pharmaceutical and cosmetic industry and also as a food flavoring agent.

**Almond Paste**—A paste made by cooking sweet and bitter almonds which have been ground and blanched in combination with sugar. It consists approximately of two parts almond to one part sugar. It is used in pastries and cakes.

**Alpha-Tocopherol**—*See Tocopherol*.

**Alum**—A preservative, the inclusive term for several aluminum-type compounds such as aluminum sulfate and aluminum potassium sulfate. It is used with EDTA to prevent discoloration of potatoes and to maintain the firmness of shrimp packs. It is also used in pickles and pickle relish.

**Aluminum Ammonium Sulfate**—A general purpose food additive that functions as a buffer and neutralizing agent. Its solubility is 1 g in 7 ml of water at 25°C and 1 g in 0.3 ml of boiling water. It is used in baking powders.

**Aluminum Calcium Silicate**—An anticaking agent used in vanilla powder. It is also used in salt up to 2 percent.

**Aluminum Nicotinate**—The aluminum salt of nicotinic acid. It is a source of niacin in foods of special dietary use.

**Aluminum Oleate**—The aluminum salt of oleic acid which is used as a binder, emulsifier, and anticaking agent. It is practially insoluble in water.

**Aluminum Sodium Sulfate**—A general purpose food additive that functions as a buffer, neutralizing agent, and firming agent. It is anhydrous and slowly soluble in water. The dodecahydrate form is readily soluble in water. It is also termed soda alum.

**Aluminum Sulfate**—A starch modifier and firming agent. The anhydrous form has a slow rate of solution while the hydrate form has a solubility of 1 g in aproximately 2 ml of water and a 1 percent solution pH of approximately 3.5. It is used up to 2 percent in combination with not more than 2 percent of 1-octenyl succinic anhydride. It is used as a firming agent in pickle and vegetable processing and as a processing aid in baked goods, gelatins, and puddings.

**Amidated Pectin**—The low-methoxyl pectin that results when some of the methoxyl groups are transformed into amide groups during deesterification with ammonia. These pectins function best in applications between 30 and 65 percent soluble solids content and pH 3.0 to 4.5. They usually require no more calcium ions than are already present in the fruit to obtain gelation. The gels formed are heat reversible. Below pH 3.5, the gels are rigid and rubbery; above pH 3.5, the gels are spreadable. Applications include flans and tart glazing. *See Pectin*.

**Amino Acids**—The food additive amino acids may be safely used as nutrients added to foods. The food additive consists of one or more of the following individual amino acids in the free, hydrated, or anhydrous form or as the hydrochloride, sodium, or potassium salts: L-Alanine, L-Arginine, L-Asparagine, L-Asparatic acid, L-Cysteine, L-Cystine, L-Glutamic acid, L-Glutamine, Aminoacetic acid (glycine),

L-Histidine, L-Isoleucine, L-Leucine, L-Lysine, DL-Methionine (not for infant foods), L-Methionine, L-Phenylalanine, L-Proline, L-Serine, L-Threonine, L-Tryptophan, L-Tyrosine, or L-Valine. The additive(s) is used to significantly improve the biological quality of the total protein in a food containing naturally occurring, primarily intact protein that is considered a significant dietary protein source. The amount of the additive added for nutritive purposes plus the amount naturally present in free and combined (as protein) form should not exceed the levels of amino acids expressed as percent by weight of the total protein of the finished food.

**Amioca—See Waxy Maize Starch.**

**Ammoniated Glycyrrhizin—See Glycyrrhizin.**

**Ammonium Alginate—**A gum that is the ammonium salt of alginic acid. It is cold water soluble and forms viscous solutions. It functions as a stabilizer and thickener and its uses include bakery fillings, gravies, and toppings.

**Ammonium Bicarbonate—**A dough strengthener, a leavening agent, a pH control agent, and a texturizer. Prepared by reacting gaseous carbon dioxide with aqueous ammonia. Crystals of ammonium bicarbonate are precipitated from solution and subsequently washed and dried.

**Ammonium Carbonate—**A dough strengthener, a leavening agent, a pH control agent, and a texturizer. It is prepared by the sublimation of a mixture of ammonium sulfate and calcium carbonate, and occurs as a white powder or a hard, white translucent mass.

**Ammonium Caseinate—**The ammonium salt of casein. It has a high nutritional value and low sodium content and is used in foods and pharmaceuticals. **See Caseinates.**

**Ammonium Chloride—**A dough conditioner and yeast food that exists as colorless crystals or white crystalline powder. Approximately 30 to 38 g dissolve in water at 25°C. The pH of a 1 percent solution at 25°C is 5.2. It is used as a dough strengthener and flavor enhancer in baked goods and as a nitrogen source for yeast fermentation. It is also used in condiments and relishes. Another term for the salt is ammonium muriate.

**Ammonium Glutamate—See Monoammonium L-Glutamate.**

**Ammonium Hydroxide**—An alkaline that is a clear, colorless solution of ammonia which is used as a leavening agent, a pH control agent, and a surface finishing agent. It is used in baked goods, cheese, puddings, processed fruits, and the production of caramels.

**Ammonium Muriate**—*See Ammonium Chloride.*

**Ammonium Persulfate**—A bleaching agent for food starch that is used at up to 0.075 percent and with sulfur dioxide up to 0.05 percent.

**Ammonium Phosphate Dibasic**—A general purpose food additive that is readily soluble in water, with approximately 57 g dissolving in 100 g of water at 0°C. A 1 percent solution has a pH of 7.6 to 8.2. It is used as a dough strengthener, firming agent, leavening agent, and pH control agent. Its uses include baked goods, alcoholic beverages, condiments, and puddings. In bakery products, up to 0.25 part per 100 parts by weight of flour is used.

**Ammonium Phosphate Monobasic**—A general purpose food additive which is readily soluble in water. A 1 percent solution has a pH of 4.3 to 5.0. It is used as a dough strengthener and leavening agent in baked goods and as a firming agent and pH control agent in condiments and puddings. It is also used in baking powder with sodium bicarbonate and as a yeast food.

**Ammonium Sulfate**—A dough conditioner, firming agent, and processing aid which is readily soluble in water with a solubility of approximately 70 g in 100 g of water at 0°C. The pH of a 0.1 molar solution in water is approximately 5.5. It is used in caramel production and as a source of nitrogen for yeast fermentation. In bakery products, up to 0.25 part per 100 parts by weight of flour is used.

**Ammonium Sulfite**—An additive used in the production of caramel.

**Amylcinnamaldehyde**—A flavoring agent that is a yellow liquid with an odor similar to jasmine. It is insoluble in glycerine and propylene, soluble in fixed oils and mineral oil. It is obtained by chemical synthesis. It can be used alone or in combination with other flavoring substances or adjuvants. It is also termed amylcinnomaldehyde.

**Anhydrous Milkfat**—*See Butter Oil.*

**Anise**—A spice that is the dried, ripe fruit of *Pimpinella anisum*, a small herb. The flavor is similar to fennel or licorice while the seed resembles caraway seed. It is used in beverages, soups, candy, liquors, and sweet pastries.

**Anise Aldehyde**—*See p-Methoxybenzaldehyde.*

**Anisyl Butyrate**—A synthetic flavoring agent that is a stable, colorless liquid of sweet cassic odor. It should be stored in glass or tin containers. It will intensify vanilla flavor and is used as a fixative. It is used in ice cream, candy, and baked goods at 5 to 15 parts per million.

**Anisyl Formate**—A synthetic flavoring agent that is a fairly stable, colorless to light yellow liquid of floral odor. It should be stored in glass, tin, or resin-lined containers. It is used in berry flavors for applications in beverages, candy, and baked goods at 3 to 10 parts per million.

**Anisyl Propionate**—A synthetic flavoring agent this is a stable, colorless liquid with a heliotrope odor. It should be stored in glass or tin containers. It is used in small concentrations to intensify vanilla, plum, and quince flavor for applications in beverages, baked goods, and candy at 6 to 20 parts per million.

**Annatto**—A color source of yellowish to reddish-orange color obtained from the seed coating of the tree *Bixa orellanna*. The oil-soluble annatto consists mainly of bixin, a carotenoid soluble in fats and oils with the color which is produced in the fat portion of the food. It has a yellow hue, very good oxidation stability, fair light stability, and good heat stability, but it is unstable above 125°C. The water-soluble annatto is norbixin (the product resulting when bixin is saponified and the methylethyl group is split off) which is dissolved as a potassium salt in lye. It is readily soluble in aqueous alkalis with the coloring occurring in the protein and starch fraction of the food. It has a yellow to orange hue and precipitates in most acid foods. The usage level is 0.5 to 10 parts per million in the finished food. It is used in sausage casings, oleomargarine, shortening, and cheese.

**Annatto Extract**—*See Bixin.*

**Anticaking Agents and Free-Flow Agents**—Substances added to finely powdered or crystalline food products to prevent caking, lumping, or agglomeration. Agents include calcium silicate, iron ammonium citrate, silicon dioxide, and yellow prussiate of soda.

**Antimicrobial Agents—*See Preservatives*.**

**Antioxidants**—Substances used to preserve food by retarding deterioration, rancidity, or discoloration due to oxidation. The most commonly used antioxidant formulations contain combinations of BHA (butylated hydroxyanisole), BHT (butylated hydroxytoluene), and propyl gallate. Natural antioxidants such as tocopherols and guaiac gum usually lack the potency of BHA, BHT, and propyl gallate combinations. Antioxidants are effective at low concentrations, that is, 0.02 percent or less.

**Apple Vinegar—*See Cider Vinegar*.**

**Arabic**—A gum obtained from breaks or wounds in the bark of *Acacia* trees. It dissolves in hot or cold water forming clear solutions which can be up to 50 percent gum acacia. The solubility in water increases with temperature. It is used in confectionary glazes to retard or prevent sugar crystallization and acts as an emulsifier to prevent fat from forming an oxidizable, greasy film. It functions as a flavor fixative in spray-drying to form a thin film around the flavor particle. It also functions as an emulsifier in flavor emulsions, as a cloud agent in beverages, and as a form stabilizer. It is also termed acacia.

**Arabinogalactan**—A gum, being the plant extract obtained from larch trees. It is soluble in hot and cold water, the water solutions up to 60 percent being fluid and above 60 percent forming a thick paste. It is stable over a wide pH range and is relatively unaffected by electrolytes. Its limited uses include dressings and pudding mixes. It is also termed larch gum.

**Arginine**—A nonessential amino acid that exists as white crystals or powder. It is soluble in water. It is used to improve the biological quality of the total protein in a food which contains naturally occurring primarily intact proteins and as a nutrient and dietary supplement.

**Arrowroot**—A starch obtained from *Mananta arundinacea*, a perennial that produces starchy rhizomes. It is neutral in flavor and of clear color. It is used as a thickener, using one-third to one-half as much as the flour or cornstarch level. It is used in fruit sauce, pie fillings, and puddings.

**Artificial Coloring**—*See Colors and Coloring Adjuncts.*

**Artificial Flavors**—*See Flavoring Agents and Adjuvants.*

**Ascorbic Acid**—It is termed vitamin C, a water-soluble vitamin that prevents scurvy, helps maintain the body's resistance to infection, and is essential for healthy bones and teeth. It is the most easily destroyed vitamin and processing is recommended in stainless steel or glass. Storage at below –18°C is recommended. In its dry form it is nonreactive, but in solution it readily reacts with atmospheric oxygen and other oxidizing agents. One part ascorbic acid is equivalent to one part erythorbic acid. It is used as a vitamin supplement in beverages, potato flakes, and breakfast foods; and as a dough conditioning agent to strengthen and condition bread roll doughs. It is also used as an antioxidant to increase shelf life in canned and frozen processed foods. It is used in conjunction with BHA, BHT, and propyl gallate to regenerate them following the chemical changes they undergo when they prevent fat rancidity in bologna and other meats. Other forms of ascorbic acid are isoascorbic (erythorbic) acid, sodium ascorbate, and sodium isoascorbate.

**Ascorbyl Palmitate**—An antioxidant formed by combining ascorbic acid with palmitic acid. Ascorbic acid is not fat soluble but ascorbyl palmitate is, thus combining them produces a fat-soluble antioxidant. It exists as a white or yellowish white powder of citric-like odor. It is used as a preservative for natural oils, edible oils, colors, and other substances. It acts synergistically with alpha-tocopherol in oils/fats. It is used in peanut oil at a maximum level of 200 mg/kg individually or in combination.

**Aspartame**—A high intensity sweetener that is a dipeptide, providing 4 calories per gram. It is synthesized by combining the methyl ester of phenylalanine with aspartic acid, forming the compound N-L-alpha-aspartyl-L-phenylalanine-1-methyl ester. It is approximately 200 times as sweet as sucrose and tastes similar to sugar. It is comparatively sweeter at low usage levels and at room temperature. Its minimum solubility is at pH 5.2, its isoelectric point. Its maximum solubility is at pH 2.2. It has a solubility of 1 percent in water at 25°C. The solubility increases with temperature. Aspartame has a certain instability in liquid systems which results in a decrease in sweetness. It decomposes to aspartylphenylalanine or to diketropiperazine (DKP) and neither of these forms is sweet. The stability of aspartame is a function of time, temperature, pH, and water activity. Maximum stability is at approximately pH 4.3. It is not usually used in baked goods because it breaks down at the high baking temperatures. It contains phenylalanine, which restricts its use for those afflicted with phenylketonuria, the inability to metabolize phenylalanine. Uses include cold breakfast cereals, desserts, topping mixes, chewing gum, beverages, and frozen desserts. The usage level ranges from 0.01 to 0.02 percent.

**Aspartic Acid**—A nonessential amino acid that exists as colorless or white crystals of acid taste. It is slightly soluble in water. It functions to improve the biological quality of a total protein in a food containing naturally occurring primarily intact protein and as a nutrient and dietary supplement.

**Azodicarbonamide**—A dough conditioner that exists as a yellow to orange-red crystalline powder practically insoluble in water. It is used in aging and bleaching cereal flour to produce a more manageable dough and a lighter, more voluminous loaf of bread. It is used in bread flours and bread as a dough conditioner. It can be used with the oxidizing agent potassium bromate. A typical use level is less than 45 parts per million.

**Babassu Oil**—The oil obtained from the nut of the babassu palm, which is native to Brazil. It is similar to coconut oil and acts as a substitute, being used in vegetable fat–based products.

**Baker's Yeast Extract**—A flavoring agent resulting from concentration of the solubles of mechanically ruptured cells of a selected strain of yeast, *Saccharomyces cerevisiae*. It may be concentrated or dried. It is used at a level not to exceed 5 percent in food.

**Baker's Yeast Glycan**—The dried cell walls of yeast, *Saccharomyces cerevisiae*, obtained from brewing. Bakers' yeast glycan is used as an emulsifier and thickener in salad dressing.

**Baker's Yeast Supplement**—A nutrient supplement which is the insoluble proteinaceous material remaining after the mechanical rupture of yeast cells of *Saccharomyces cerevisiae* and removal of whole cell walls by centrifugation and separation of soluble cellular materials.

**Baking Powder**—A leavening agent that consists of a mixture of sodium bicarbonate, one or more leavening agents such as sodium aluminum phosphate or monocalcium phosphate, and an inert material such as starch. The inert material keeps the reactive components physically separated and minimizes premature reaction. It should yield not less than 12 percent of available carbon dioxide.

**Baking Soda—*See Sodium Bicarbonate*.**

**Balsam Peru Oil**—A flavoring agent, which is liquid, and yellow to pale green in color. It is viscous and has a sweet balsamic odor. It is insoluble in glycerin, slightly soluble in propylene glycol, soluble and turbid in mineral oil, and soluble in fixed oils. It is obtained by extraction or distillation of Peruvian Balsam obtained from *Myroxylon pereirae* Royal Klotsche of the *Leguminosae* family. It can be used alone or in combination with other flavoring substances or adjuvants.

**Barley**—A cereal grain of which there are winter and spring types. It is used in malting (the conversion of grain to malt used in beer production) as malted barley. Malt flour is used in baking, cereals, and sauces. Pearled barley, in which the hull and outer kernel are removed by abrasive action, is found in barley soups. Barley flour and flakes are used in baked products. Barley is high in carbohydrates and contains protein, calcium, phosphorus, and B vitamins.

**Barley, Malted**—*See Malted Barley.*

**Basil**—A spice obtained from the dried leaves and tender stems of *Ocimum basilicum* L. The fresh basil resembles licorice in flavor and the dried leaves have a lemony anise-like quality. This delicate herb can be used generously and has an affinity for tomato-based products. It is used in tomato-based recipes, with vegetables, and in tomato sauce. It is also termed sweet basil.

**Bay Leaves**—A spice flavoring that consists of the dried leaves obtained from the evergreen tree *Laurus nobilis*, also called sweet bay or laurel tree. They have a sweet, herbaceous flavor and are used as an herb. They are aromatic when crushed and find use in meat, soup, and stew.

**Beeswax**—The purified wax obtained from the honeycomb of the bee is insoluble in water and is sparingly insoluble in cold alcohol. It is used to glaze candy, in chewing gum, in confections, and as a flavoring agent.

**Beet Extract**—A natural red colorant obtained from beets that is of very good water solubility and has fair pH stability, poor heat stability, and good light stability. It is colored by betacyanins which include red and yellow compounds, the major red pigment being betanin. The betanin accounts for 75 to 95 percent of the total pigment content. It is available in concentrate and powder forms and is used in yogurt, beverages, candies, and desserts.

**Beets, Dehydrated**—*See Beet Extract.*

**Beet Sugar**—*See Sugar.*

**Benne**—*See Sesame Seed.*

**Bentonite**—A general purpose additive that is used as a pigment and colorant and to clarify and stabilize wine.

**Benzaldehyde**—A flavoring agent which is liquid and colorless, and has an almond-like odor. It has a hot (burning) taste. It is oxidized to benzoic acid when exposed to air and deteriorates under light. It is miscible in volatile oils, fixed oils, ether, and alcohol; it is sparingly soluble in water. It is obtained by chemical synthesis and by natural occurrence in oils of bitter almond, peach, and apricot kernel. It is also termed benzoic aldehyde.

**Benzoate of Soda**—*See Benzoic Acid*.

**Benzoic Acid**—A preservative that occurs naturally in some foods such as cranberries, prunes, and cinnamon. It is most often used in the form of sodium benzoate because of the low aqueous solubility of the free acid. Sodium benzoate is 180 times as soluble in water at 25°C as benzoic acid. The salt in solution is converted to the acid which is the active form. The optimum pH range for microbial inhibition is pH 2.5 to 4.0. It is used in acid foods such as carbonated beverages, fruit juices, and pickles. It is also termed benzoate of soda.

**Benzoic Aldehyde**—*See Benzaldehyde*.

**Benzoyl Peroxide**—A colorless, crystalline solid with a faint odor of benzaldehyde resulting from the interraction of benzoyl chloride and a cooled sodium peroxide solution. It is insoluble in water. It is used in specified cheeses at 0.0002 percent of milk level. It is used for the bleaching of flour, slowly decomposing to exert its full bleaching action, which results in whiter flour and bread.

**Benzyl Propionate**—A flavoring agent which is liquid, colorless and has a sweet, fruity odor. It is soluble in most fixed oils and alcohol, slightly soluble in propylene glycol, and insoluble in glycerin.

**Beta-Apo-8'-Carotenal**—A colorant that is a carotenoid producing a light to dark orange hue. It has fair light stability, poor oxidation stability, and good pH stability. It is insoluble in water but is available in water-dispersible, oil-dispersible, and oil-soluble forms. It has vitamin A activity. It is used in cheese and cheese sauces, and dressings. The maximum usage level is 33 parts per million. Related colorants are beta-carotene and canthaxanthin.

**Beta-Carotene**—A colorant that is a carotenoid producing a yellow to orange hue. It has good tinctorial strength, fair light stability, poor oxidation stability, and good pH stability. It is insoluble in water but is available in water-dispersible, oil-dispersible, and oil-soluble forms. It has vitamin A activity. It has a natural resistance to ascorbic acid reduction in beverages and thus is used in orange-colored liquid products. It is used in margarine, oils, cheese, and puddings at levels required to produce the desired color. Related colorants are canthaxanthin and beta-apo-8'-carotenal.

**Bicarbonate of Soda—*See Sodium Bicarbonate*.**

**Biotin**—A water-soluble vitamin that is a nutrient and dietary supplement. It is relatively stable to heat and storage and is found in eggs, liver, peanuts, milk, and meat. It functions in the metabolism of carbohydrates, proteins, and fats. It is essential for the activity of many enzyme systems.

**Birch**—An artificial flavoring used in soft drinks such as birch beer.

**Bixin**—A carotenoid that is the main coloring component of annatto. It is obtained from the *Bixa orellana* tree. Bixin is soluble in fats and oils and the produced color is found in the fat fraction of the food. It has a yellow hue, very good oxidation stability, fair light stability, and good heat stability, but it is poor at very high temperatures, such as above 125°C. One part bixin is equivalent to 1.5 parts carotene. It is used at 0.5 to 10 parts per million in finished foods, such as margarine, salad dressings, popcorn oil, and baked goods. It is also termed annatto extract. *See Annatto*.

**Bleached Flour**—Flour that has been whitened by the removal of the yellow pigment. The bleaching can be obtained during the natural aging of the flour or can be accelerated by chemicals that are usually oxidizing agents which oxidize the carotenoid pigments to a nearly colorless product. The oxidizing agents also improve the flour performance by their effect on the protein. The process improves the baking quality by allowing the formation of high ratio cakes that would be likely to collapse if prepared with untreated flour.

**Bleaching Agents—*See Flour Treating Agents*.**

**Bodying Agent—*See Stabilizers and Thickeners*.**

**Bran**—The seed husks or outer coatings of cereals, such as wheat, rye, and oats, that are separated from the flour. It is used in bran flakes cereal.

**Bread Flour**—A hard-wheat flour, which generally contains in excess of 10.5 percent protein and is obtained from straight or long patent flours. These flours have high absorption and good mixing tolerance which make them suitable in yeast-leavened breads.

**Brilliant Blue FCF—*See FD&C Blue #1*.**

**Bromated Flour**—A white flour to which potassium bromate is added at a level not to exceed 50 parts per million. It is used in baked goods.

**Brominated Vegetable Oil**—(BVO) A vegetable oil whose density has been increased to that of water by combination with bromine. Flavoring oils are dissolved in the brominated oil which can then be added to fruit drinks. The adjustment of the specific gravity makes it possible to obtain stable finished beverages. If the oil phase gravity is too low the emulsion will form a ring, and if it is too high a white precipitate may form. It is also used in formulating cloud agents. Its use is limited to 15 parts per million.

**Brown Sugar**—A sweetener that consists of sucrose crystals covered with a film of cane molasses. Molasses gives it the characteristic color and flavor. There are three grades: light, medium, and dark, which vary in sucrose content and color. It is used in baked goods, glazes, toppings, and fillings.

**Bulgur**—A precooked cracked wheat that retains the bran and germ fraction of the grain. It resembles whole wheat nutritionally and is sometimes termed parboiled wheat. It is an excellent source of whole grain, protein, and carbohydrates. It is reconstituted by cooking or soaking in liquid. It can be used in bread, casseroles, and salads, or can be eaten as such.

**Bulgur Wheat**—Whole wheat kernels that are cleaned, cooked (parboiled), dried, ground, and sifted for sizing. The resulting wheat has a nut-like flavor, extended shelf life, and is higher in most nutrients than rice. It can impart texture, water absorption, fiber, and nutrients. It is used in taboule salads, pilaf, soups, stuffings, and bakery products.

**Bulking Agents—*See Stabilizers and Thickeners*.**

**Butter, Clarified**—Butter that has undergone purification by the removal of solid particles or impurities that may affect the color, odor, or taste.

**Butter Fat—*See Milkfat*.**

**Buttermilk**—The product that remains when fat is removed from milk or cream in the process of churning into butter. Cultured buttermilk is prepared by souring buttermilk, or more commonly skim milk, with a suitable culture that produces a desirable taste and aroma. It is used as a beverage, as an ingredient in baked goods, and in dressings.

**Buttermilk, Dried**—The powder form of buttermilk. It is similar in composition to nonfat dry milk but of higher fat concentration, much of which is phospholipids which provide good emulsifying and whipping properties. It is used in dry mix, desserts, soups, and sauces.

**Butter Oil**—The clarified fat portion of milk, cream, or butter obtained by the removal of the nonfat constituent. It contains not less than 99.7 percent milkfat, not more than 0.2 percent moisture, and not more than 0.05 percent milk solids nonfat. It is used in frozen desserts, puddings, and syrups. It is also termed anhydrous milkfat, or ghee.

**Butyl Acetate**—A flavoring agent which is a clear, colorless liquid possessing a fruity and strong odor. It is sparingly soluble in water and miscible in alcohol, ether, and propylene glycol. It is also termed *n*-butyl acetate.

**Butylated Hydroxyanisole**—(BHA) An antioxidant that imparts stability to fats and oils and should be added before oxidation has started. It is a mixture of 3-tert-butyl-4-hydroxyanisole and 2-tert-butyl-4-hydroxyanisole. In direct addition, the fat or oil is heated to 60 to 70°C and the BHA is added slowly under vigorous agitation. The maximum concentration is 0.02 percent based on the weight of the fat or oil. It may protect the fat-soluble vitamins A, D, and E. It is used singly or in combination with other antioxidants. It is used in cereals, edible fat, vegetable oil, confectionary products, and rice.

**Butylated Hydroxytoluene**—(BHT) An antioxidant that functions similarly to butylated hydroxyanisole (BHA) but is less stable at high temperatures. It is also termed 2,6-di-tert-butyl-para-cresol. *See Butylated Hydroxyanisole.*

**Butyl Butyryllactate**—A synthetic flavoring agent that is a stable, colorless to light yellow liquid with the odor of cooked butter. It is miscible with alcohol and most fixed oils, soluble in propylene glycol, and insoluble in glycerine and water. It should be stored in glass, tin, or resin-lined containers. It is used in butter flavors with applications in baked goods and candy at 14 to 60 parts per million.

**Butyl Heptanoate**—A synthetic flavoring agent that is a stable, colorless liquid of fruity odor. It is stored in glass or tin containers.

It is used in flavors such as apple, blackberry, and ginger beer with applications in candy and baked goods at 25 parts per million.

**Butylhydroquinone**—*See Tertiary Butylhydroquinone.*

**Butylparaben**—*See Parabens.*

**Butyric Acid**—A fatty acid that is commonly obtained from butter fat. It has an objectionable odor which limits its uses as a food acidulant or antimycotic. It is an important chemical reactant in the manufacture of synthetic flavoring, shortening, and other edible food additives. In butter fat, the liberation of butyric acid which occurs during hydrolytic rancidity makes the butter fat unusable. It is used in soy milk–type drinks and candies.

# C

**Cacao Butter**—*See Cocoa Butter*.

**Caffeine**—A white powder or needles that are odorless and have a bitter taste. It occurs naturally in tea leaves, coffee, cocoa, and cola nuts. It is a food additive used in soft drinks for its mildly stimulating effect and distinctive taste note. It is used in cola-type beverages and is optional in other soft drinks up to 0.02 percent.

**Cake Flour**—A soft wheat flour that is generally a short patented flour containing less than 10 percent protein. Such flours are low in water absorption and are of short mixing time and tolerance. It is used in chemically leavened cakes, cookies, and pastries.

**Calciferol**—A fat-soluble vitamin, termed vitamin $D_2$, which is stable unless oxidized. It is necessary for growth and maintenance of teeth and bones and the normal utilization of calcium and phosphorus; it is used medicinally in the treatment of rickets and as a dietary supplement. Its sources include fish liver and vitamin D–fortified milk.

**Calcium**—An alkaline earth element that contributes toward bone and teeth formation, muscle contraction, and blood clotting. It occurs in milk, vegetables, and egg yolk.

**Calcium Acetate**—The calcium salt of acetic acid which functions as a sequestrant and mold control agent. It contains approximately 25 percent calcium. It is a white odorless powder which is readily soluble in water with a solubility of approximately 37 g in 100 g water at 0°C. Its solubility decreases with increasing temperature, with a solubility of approximately 29 g in 100 g of water at 100°C.

**Calcium Acid Phosphate**—*See Monocalcium Phosphate*.

**Calcium Alginate**—The calcium salt of alginic acid which functions as a stabilizer and thickener. The partial obtainment of calcium alginate by the reaction of the water-soluble alginate with calcium ions is used to obtain viscosity and gel formation. It is used in icings, imitation pulp, dessert gels, and fabricated fruits.

**Calcium Ascorbate**—The salt of ascorbic acid which is a white to slightly yellow crystalline powder. It is soluble in water and the pH of a 10 percent solution is 6.8 to 7.4. It functions as an antioxidant and preservative. *See Ascorbic Acid.*

**Calcium Biphosphate**—*See Monocalcium Phosphate.*

**Calcium Bromate**—A dough conditioner and maturing and bleaching agent which exists as a white crystalline powder. It is very soluble in water and is used in flour and dough.

**Calcium Carbonate**—The calcium salt of carbonic acid which is used as an anticaking agent and dough strengthener. It is available in varying particle sizes ranging from coarse to fine powder. It is practically insoluble in water and alcohol, but the presence of any ammonium salt or carbon dioxide increases its solubility while the presence of any alkali hydroxide reduces its solubility. It has a pH of 9 to 9.5. It is the primary source of lime (calcium oxide) which is made by heating limestone in a furnace. Calcium carbonate is used as a filler in baking powder, for calcium enrichment, as a mild buffering agent in doughs, as a source of calcium ions in dry mix desserts, and as a neutralizer in antacids. It is also termed limestone.

**Calcium Carrageenan**—*See Carrageenan.*

**Calcium Caseinate**—The calcium salt of casein. Properties include low viscosity, settling out of water, opaqueness, no heat stability, and chalky texture. It contains the range of essential amino acids present in sodium caseinate but has a higher concentration of calcium. It is useful in applications requiring low absorption properties. It is used as a protein source in imitation cheese, and in special diet foods to replace sodium caseinate where sodium must be restricted. It is used to improve the whipping properties of vegetable whipped toppings, and as a binder. *See Caseinates.*

**Calcium Chloride**—A general purpose food additive, the anhydrous form being readily soluble in water with a solubility of 59 g in 100 ml of water at 0°C. It dissolves with the liberation of heat. It also exists as calcium chloride dihydrate, being very soluble in water with a solubility of 97 g in 100 ml at 0°C. It is used as a firming agent for canned tomatoes, potatoes, and apple slices. In evaporated milk, it is used at levels not more than 0.1 percent to adjust the salt balance so as to prevent coagulation of milk during sterilization. It is used with disodium EDTA to protect the flavor in pickles and as a source of calcium ions for reaction with alginates to form gels.

**Calcium Citrate**—The calcium salt of citric acid which functions as a sequestrant, buffer, and firming agent. It is a white, odorless powder which is slightly soluble in water. It is used as a firming agent for peppers and lima beans and is used to improve the baking properties of flour.

**Calcium Diacetate**—The salt of acetic acid which is used as a preservative and sequestrant.

**Calcium Disodium EDTA**—*See Disodium Calcium EDTA*.

**Calcium Gluconate**—A white crystalline granule or powder that functions as a firming agent, formulation aid, sequestrant, and stabilizer. At room temperature the anhydrous form has a solubility of approximately 1 g in 30 ml of water, which improves in boiling water to approximately 1 g in 5 ml of water. It also exists as calcium gluconate (monohydrate). It is used as a source of calcium ions for sodium alginate gels, and as a calcium fortifier in baked goods, puddings, and dairy product analogs. It functions as a coagulation aid in milk and instant pudding powders and as a means of masking the bitter aftertaste of some artificial sweeteners.

**Calcium Glycerophosphate**—A nutrient and dietary supplement which is a white odorless powder of poor water solubility. It is used in dental impression material and baking powder.

**Calcium Hydrate**—*See Calcium Hydroxide*.

**Calcium Hydroxide**—A general food additive made by adding water to calcium oxide (lime). It has poor water solubility with a solubility of 0.185 g in 100 g water at 0°C. The pH of a water solution at 25°C is approximately 12.4. It is used to promote dispersion of ingredients in sauces, creamed spinach, and a frozen pea/potato dish. It is used at 0.1 percent to stabilize the potassium iodide of iodized salt, and it is used as a neutralizer for soured cream prior to buttermaking. It is also termed hydrated lime, calcium hydrate, and slaked lime.

**Calcium Iodate**—A source of iodine that is a white powder of slight solubility in water, but greater solubility in water containing iodides or amino acids. It is more stable than the iodide form. It is used as a dough conditioner in bread and is a source of iodine in table salts.

**Calcium Lactate**—The calcium salt of lactic acid which is soluble in water. It has a solubility of 3.4 g per 100 g of water at 20°C and is very soluble in hot water. It is available as a monohydrate, trihydrate, and pentahydrate. The trihydrate and pentahydrate have solubilities of

9 g in 100 ml water at 25°C. It contains approximately 14 percent calcium. It is used to stabilize and improve the texture of canned fruits and vegetables by converting the labile pectin to the less soluble calcium pectate. It thereby prevents structural collapse during cooking. It is used in angel food cake, whipped toppings, and meringues to increase protein extensibility which results in an increase of foam volume. It is also used in calcium fortified foods such as infant foods and is used to improve the properties of dry milk powder.

**Calcium Lactobionate**—The calcium salt of lactobionic acid (4-(B, D-galactosido)-D-gluconic acid) produced by the oxidation of lactose. It is soluble in water and is used as a firming agent in dry pudding mixes.

**Calcium Oxide**—A general food additive consisting of white granules or powder of poor water solubility. It is obtained by heating limestone (calcium carbonate) in a furnace. It is also termed lime or quicklime. It is used as an anticaking agent, firming agent, and nutritive supplement in applications such as grain products and soft candy.

**Calcium Pantothenate**—A nutrient and dietary supplement which is the calcium chloride double salt of calcium pantothenate. It is a white powder of bitter taste and has a solubility of 1 g in 3 ml of water. It is used in special dietary foods.

**Calcium Pectinate**—The salt of pectin which is obtained from citrus or apple fruit. It results from the interaction of low-methoxyl pectin with calcium ions to form a gel. It is used as a gel coating for meat products and to form food gels. *See Low-Methoxyl Pectin*.

**Calcium Peroxide**—A dough conditioner which exists as a white or yellowish powder or granule that is insoluble in water. It improves dough strength, grain, and texture, and increases absorption and crumb resiliency. It is used in bakery products.

**Calcium Phosphate**—A compound existing in several forms which include the monobasic, dibasic, and tribasic forms of calcium phosphate. As calcium phosphate monobasic, also termed monocalcium phosphate, calcium biphosphate, and acid calcium phosphate, it is used as a leavening agent and acidulant. Calcium phosphate dibasic, also termed dicalcium phosphate dihydrate, is used as a dough conditioner and mineral supplement. Calcium phosphate tribasic, also termed tricalcium phosphate and precipitated calcium phosphate, is used as an anticaking agent, mineral supplement, and conditioning agent.

**Calcium Phosphate, Dibasic Anhydrous**—*See Dicalcium Phosphate, Anhydrous*.

**Calcium Phosphate, Dibasic Hydrous**—*See Dicalcium Phosphate, Dihydrate*.

**Calcium Phosphate Monobasic**—*See Monocalcium Phosphate*.

**Calcium Phosphate Tribasic**—*See Tricalcium Phosphate*.

**Calcium Propionate**—The salt of propionic acid which functions as a preservative. It is effective against mold, has limited activity against bacteria, and no activity against yeast. It is soluble in water with a solubility of 49 g per 100 ml of water at 0°C and insoluble in alcohol. It is less soluble than sodium propionate. Its optimum effectiveness is up to pH 5.0 and it has reduced action above pH 6.0. It is used in bakery products, breads, and pizza crust to protect against mold and "rope." It is also used in cold-pack cheese food and pie fillings. Typical usage level is 0.2 to 0.3 percent and 0.1 to 0.4 percent based on flour weight.

**Calcium Pyrophosphate**—A nutrient and dietary supplement that exists as a white odorless powder, insoluble in water. It is used in dental impression materials and as a buffer.

**Calcium Saccharin**—A sweetener that is the calcium form of saccharin, existing as white crystals of powder with a solubility of 1 g in 1.5 ml of water. Sodium saccharin is the more common form, but calcium saccharin is available for nonsodium diets. In this form it is about 500 times as sweet as sucrose. *See Saccharin*.

**Calcium Silicate**—An anticaking agent that exists in different forms, which are insoluble in water. It is used in salt to enhance flowability under extremely high humidity conditions. It is also used in baking powder and fabricated chips to absorb water or other liquids.

**Calcium Sorbate**—A preservative that is the calcium salt of sorbic acid. It is not the most common form. Its solubility in water or fat is very limited and therefore it is used on surfaces for preservation. It is permitted in cheese and wrapping materials.

**Calcium Stearate**—The calcium salt of stearic acid which functions as an anticaking agent, binder, and emulsifier. It is used in garlic salt, dry molasses, vanilla and vanilla-vanillin powder, salad dressing mix, and meat tenderizer. It can be used for mold release in the tableting of pressed candies.

**Calcium Stearyl-2-Lactylate**—A mixture of calcium salts of stearyl lactylic acids and minor proportions of other calcium salts of related acids. It is manufactured by the reaction of stearic acid and lactic acid and conversion to the calcium salts, and is used as follows: as a dough conditioner in yeast-leavened bakery products and prepared mixes for yeast-leavened bakery products in an amount not to exceed 0.5 part for each 100 parts by weight of flour used; as a whipping agent in liquid and frozen egg white at a level not to exceed 0.05 percent; in whipped vegetable oil topping at a level not to exceed 0.3 percent of the weight of the finished whipped vegetable oil topping; and as a conditioning agent in dehydrated potatoes in an amount not to exceed 0.5 percent by weight.

**Calcium Sulfate**—A general additive available as both calcium sulfate anhydrous, made by the high-temperature calcining of gypsum which is then ground and separated, and calcium sulfate dihydrate, which is made by grinding and separating gypsum containing about 20 percent water of crystallization. Calcium sulfate anhydrous contains approximately 29 percent calcium, and calcium sulfate dihydrate contains approximately 23 percent calcium. It is used, among other things, as a filler and baking powder for standardization purposes; a firming agent in canned potatoes, tomatoes, carrots, lima beans, and peppers; in dough as a source of calcium ions (because the absence of calcium ions causes bread dough to be soft and sticky and to produce bread of poor quality); in soft-serve ice cream to produce dryness and stiffness; as a calcium ion source for reaction with alginates to form dessert gels; and as a calcium source for food enrichment.

**Cananga Oil**—A flavoring agent. It is a yellow liquid with a harsh, flowery odor. It is soluble in most fixed oils and mineral oil, and insoluble in glycerin and propylene glycol. It is obtained by distillation of flowers of *Cananga odorato* Hook and Thomas (tree of the *Anonaceae* family).

**Candelilla Wax**—A lubricant and surface finishing agent obtained from the candelilla plant. It is a hard, yellowish-brown, opaque-to-translucent wax. Candelilla wax is prepared by immersing the plants in boiling water containing sulfuric acid and skimming off the wax that rises to the surface. It is composed of about 50 percent hydrocarbons with smaller amounts of esters and free acids. It is used in chewing gum and hard candy.

**Cane Sugar**—*See Sugar.*

**Canola Oil**—*See Rapeseed Oil, Low Erucic Acid.*

**Canthaxanthin**—A synthetic red colorant that is the carotenoid of most intense red color. It is available in oil-soluble, oil-dispersible, and water-dispersible forms. It has fair pH, heat, light, and chemical stability with a low tinctorial strength. Unlike the carotenoids beta-carotene and beta-apo-8'-carotenal, it does not possess vitamin A activity. Maximum usage level is 66 parts per million. Uses include carbonated soft drinks, salad dressing, and spaghetti sauce.

**Caprylic Acid**—A flavoring agent considered to be a short or medium chain fatty acid. It occurs normally in various foods and is commercially prepared by oxidation of *n*-octanol or by fermentation and fractional distillation of the volatile fatty acids present. It is used in maximum levels, as served, of 0.13 percent for baked goods; 0.04 percent for cheeses; 0.005 percent for fats and oils, frozen dairy desserts, gelatins and puddings, meat products, and soft candy; 0.016 percent for snack foods; and 0.001 percent or less for all other food categories. It is also termed octanoic acid.

**Caramel**—A colorant that is an amorphous, dark brown product resulting from the controlled heat treatment of carbohydrates such as dextrose, sucrose, and malt syrup. It is available in liquid and powdered forms, providing shades of brown. In coloring a food with caramel, the food components must have the same charge as the particles of caramel, otherwise the particles will attract one another and precipitate out. Caramel can exist as several types, for example, acid-proof caramel of negative charge which is used in carbonated beverages, acidified solutions, bakers' and confectioners' caramel which are used in baked goods; and dried caramel for dry mixes. Major uses are in coloring beverages such as colas and root beers and in baked goods.

**Caraway**—A spice that is a seed obtained from the tree *Carum carvi*. It has a flavor similar to dill. It is used in rolls, bread, meats, and some cheeses.

**Carbonated Water**—A beverage made by absorbing carbon dioxide in water. The carbon dioxide influences flavor because increased carbonation increases mouth feel. Gas retention is more common in low calorie–type beverages because of the absence of sugar solids.

**Carbon Dioxide**—A gas obtained during fermentation of glucose (grain sugar) to ethyl alcohol. It is used in pressure-packed foods as a propellant or aerating agent and is also used in the carbonation of beverages. It is released as a result of the acid carbonate reaction of leavening agents in baked goods to produce an increase in volume. As a solid, it is termed dry ice and is used for freezing and chilling.

**Carboxymethylcellulose**—(CMC) A gum that is water-soluble cellulose ether manufactured by reacting sodium monochloroacetate with alkali cellulose to form sodium carboxymethylcellulose. It dissolves in hot or cold water and is fairly stable over a pH range of 5.0 to 10.0, but acidification below pH 5.0 will reduce the viscosity and stability except in a special acid-stable type of CMC. A variety of types are available which differ in viscosity and degree of substitution (the number of sodium groups per unit). It functions as a thickener, stabilizer, binder, film former, and suspending agent. It is used in a variety of foods to include dressings, ice cream, baked goods, puddings, and sauces. The usage range is from 0.05 to 0.5 percent.

**Cardamon**—A spice that is a dried, ripe seed of *Elettaria cardamomum*, a biennial plant. It has a pungent aroma and is reddish-brown in color. The flavor is sweet and spicy with a camphoraceous note. It is used in whole form to flavor hot fruit punches, pickles, and marinades. It is used in the ground form in bread, cookies, desserts, and meats.

**Carmine**—The red colorant aluminum lake of carminic acid which is the coloring pigment obtained from dried bodies of the female insect *Coccus cacti*. It is brilliant red to purplish in color, having a low tinctorial strength, and can be solubilized in ammonia. It is used in a pink color in coatings.

**Carnauba Wax**—A general purpose food additive that is a hard and brittle wax. It is obtained from the leaf buds and leaves of the Brazilian wax palm *Copernicia cerifera*. It is the hardest wax known and is used in candy glaze.

**Carob**—A cocoa substitute obtained from the pods of the carob tree *Ceratonia siliqua*. The pods are kibbled, roasted, and ground into a powder which is similar in appearance and fragrance to cocoa powder. Carob powder has less than 1 percent fat and 42 to 48 percent sugar, while cocoa powder has approximately 23 percent fat and 5 percent sugar. Cocoa does not contain any measurable amounts of fructose so the presence of carob in cocoa can be detected by the presence of fructose. It is used in candy, drinks, bakery products, and dairy applications, and as a single ingredient in health food products.

**Carob Gum—*See Locust Bean Gum*.**

**Carotene—**A colorant and provitamin, being a hydrocarbon which is one of two subgroups of the carotenoids (yellow, orange, or red pigments). The other subgroup is xanthophylls. Carotene functions as a colorant with beta-apo-8'-carotenal being a red-orange carotenoid and beta-carotene being a yellow carotenoid. It is also a vitamin A precursor that is converted by the body to vitamin A. It is used in ice cream, cheese, and other dairy products.

**Carrageenan—**A gum that is a seaweed extract obtained from red seaweed *Chondrus crispus* (also known as Irish moss), *Gigartina*, and *Eucheuma* species. *Chondrus crispus* yields kappa and lambda carrageenans. *Gigartina* yields kappa and lambda carrageenans. *Eucheuma* yields kappa and iota carrageenans. It exists as various salts or mixed salts of a sulfate ester. It is classified mainly as kappa, iota, and lambda types which differ in solubility and gelling properties. The kappa and iota types require hot water (above 71°C) for complete solubility and can form thermally reversible gels in the presence of potassium and calcium cations, respectively. The kappa gels are brittle with syneresis while the iota gels are more elastic without syneresis. The lambda type is cold-water soluble and does not form gels. Kappa and iota carrageenan are very reactive with milk protein products. Carrageenan is used to stabilize milk protein at 0.01 to 0.05 percent and to form water gels at 0.5 to 1.0 percent. Its uses include dairy products, water gel desserts, and low-calorie jellies. A typical use level in water systems is 0.2 to 1.0 percent and milk systems is 0.01 to 0.25 percent.

**Carubin—*See Locust Bean Gum*.**

**Carvacrol—**A flavoring agent that is a colorless to pale yellow liquid. It has a spicy and pungent odor, resembling thymol. It is insoluble in water and soluble in alcohol and ether. It is a mixture of the isomeric carvacrols (isopropyl o-creols), and is obtained by chemical synthesis. It is also an ingredient of savory, a fragrant herb in nature.

**Casein**—The principal milk protein which is prepared commercially from skim milk by the precipitation with lactic, hydrochloric, or sulfuric acid. It can also be produced by the use of lactic acid–producing bacteria. Caseins are usually identified according to the acid used, such as lactic acid casein, hydrochloric acid casein, and sulfuric acid casein. The principal form in which casein is used is casein salts, of which sodium and calcium caseinate are the most common. Rennet casein is obtained from skim milk by the precipitation with a rennet-type enzyme. Casein is used in the protein fortification of cereals and bread, and in fabricated cheeses.

**Caseinates**—Salts of casein that are produced by neutralizing acid casein to pH 6.7 with calcium or sodium hydroxide, producing the most common forms, which are calcium caseinate or sodium caseinate. Other forms of casein are potassium and ammonium caseinate. The caseinates provide a source of protein and function as emulsifiers, water binders, and whipping aids. The relative water absorption of casein salts is: calcium caseinate—130 percent, potassium caseinate—200 percent, sodium caseinate—250 percent. Its uses include processed meats, whipped toppings, coffee whiteners, egg substitutes, and diet foods.

**Castor Oil**—A release and antisticking agent used in hard candy production. Its concentration is not to exceed 500 parts per million. It is used in vitamin and mineral tablets, and as a component of protective coatings.

**Cayenne Pepper**—*See Pepper, Cayenne.*

**Celery Seed**—A spice made from the dried, ripe fruit of the herb *Apium graveolens*, related to the parsley family. It is used lightly so as not to dominate in flavor. It is used in sauces, salads, meats, and soups.

**Cellulose**—A carbohydrate polymer made up of glucose units. It consists of fibrous particles and is used as a fiber source and bulking agent in low-calorie formulations.

**Cheese Culture**—Bacteria used in the coagulation of the milk protein casein in the formation of cheese. It converts milk into cheese curd by the reduction of pH followed by processing to precipitate the protein as a curd.

**Cheese Powder**—A dry form of cheese prepared by slurrying cheese in water to 35 to 45 percent solids and further processing into a powder form. Cheese powders are water soluble. They are used in instant soups, dry spaghetti sauce, dry sauces, and snack foods.

**Chelating Agents**—*See Sequestrants.*

**Chervil**—A spice derived from the plant *Anthriscus cerefolium* which is related to the parsley family. It is used in soufflés, sauces, meats, and fish.

**Chewing Gum Base**—A base, containing masticatory substances such as chicle, used in the manufacture of chewing gum.

**Chicle**—A natural masticatory substance of vegetable origin which is used in chewing gum base. It is the latex of the sapodilla tree, obtained by cutting the bark to yield the latex which is boiled to remove about two-thirds of the water. The resulting semisolid mass is molded into chicle blocks which form the base for chewing gum.

**Chilte**—A substance of vegetable origin used as a masticatory substance in chewing gum base.

**Chiquibul**—A substance of vegetable origin used as a masticatory substance in chewing gum base.

**Chives**—A spice from the *Allium schoenoprasum* plant whose slender rush-like green leaves are chopped and used to provide a subtle onion flavor and to enhance food appearance. It is also used as a garnish and topping.

**Chlorine**—A gas used to age and bleach flour.

**Chlorine Dioxide**—A gas used in bleaching and aging flour. It acts on the flour almost instantly, resulting in improved color and dough properties. Because usage levels are low, the bleaching action is limited.

**Chloropentafluoroethane**—A propellant and aerating agent for foamed or sprayed foods.

**Chlorophyll**—A colorant that is a green pigment present in all green plants. It is used in sausage casings, oleomargarine, and shortening.

**Chocolate**—A solid or semiplastic food made from chocolate liquor derived from cocoa nibs, which are obtained from the cocoa bean. Chocolate contains more fat and less protein than cocoa. The products derived from chocolate include bitter or plain chocolate; sweet chocolate containing sugar, milk, flavoring, and cocoa butter; and milk chocolate, which is made from sweet or bitter chocolate plus a milk source with or without cocoa butter and flavoring. It is used as a flavor in candy, dairy products, and baked goods.

**Chocolate Liquor**—*See Cocoa Liquor.*

**Cholic Acid**—An emulsifier that exists as colorless plates or a white crystalline powder which has a bitter taste with a sweetish aftertaste. It is slightly soluble in water. It functions as an emulsifying agent in egg white.

**Choline**—A substance grouped as a member of the vitamin B complex, although not a vitamin by definition. It is water soluble and is important in nerve function and fat metabolism. It occurs in egg yolk, beef liver, and grains.

**Cider Vinegar**—The product made by the alcoholic and subsequent acetous fermentation of apple juice or concentrate thereof. It contains not less than 4 g acetic acid in 100 cm$^3$ at 20°C. It has a light to medium amber color. It is used in salad dressings, mayonnaise, and sauces. The term *vinegar* refers to cider vinegar, also termed apple vinegar.

**Cinnamic Acid**—A flavoring agent that consists of crystalline scales, white in color, with an odor resembling honey and flowers. It is slightly soluble in water, soluble in alcohol, chloroform, acetic acid, acetone, benzene, and most oils, and alkali salts soluble in water. It is obtained by chemical synthesis. It is also termed 3-phenylpropenoic acid.

**Cinnamon**—A spice made from the dried bark of the evergreen tree *Cinnamomum cassia.* Commercial types are Saigon Cassia and Batavia Cassia. Ceylon cinnamon is the dried inner bark of shoots of *C. zeylanicum* Nees. In the ground form it is used in beverages, desserts, and fruits while in the stick form it is used in beverages, meats, and fruits.

**Cinnamyl Anthranilate**—A flavoring agent that is a powder which may be red or yellow. It has an odor resembling anthranilates, fruity and characteristically balsamic. It is insoluble in water, and soluble in alcohol, chloroform, and ether. It is obtained by chemical synthesis.

**Cinnamyl Isobutyrate**—A synthetic flavoring agent that is a moderately stable, colorless to light yellow liquid of dry fruity color. It is stored in glass or tin containers. It is used to give a lift to jasmine with applications in baked goods and candy at 8 parts per million.

**Citral**—A liquid flavoring agent, light yellow in color with a citrus odor. It occurs in lemon and lemongrass oils. It is usually obtained from citral-containing oils by chemical means but may also be prepared synthetically. It is soluble in fixed oils, mineral oil, and propylene glycol. It is moderately stable and should be stored in glass, tin, or resin-lined containers. It is used in flavors for lemon with applications in candy, baked goods, and ice cream at 20 to 40 parts per million. It is also termed 2,6-dimethyl-octadian-2-6-a1-8.

**Citric Acid**—An acidulant and antioxidant produced by mold fermentation of sugar solutions and by extraction from lemon juice, lime juice, and pineapple canning residue. It is the predominant acid in oranges, lemons, and limes. It exists in anhydrous and monohydrate forms. The anhydrous form is crystallized in hot solutions and the monohydrate form is crystallized from cold (below 36.5°C) solutions. Anhydrous citric acid has a solubility of 146 g and monohydrate citric acid has a solubility of 175 g per 100 ml of distilled water at 20°C. A 1 percent solution has a pH of 2.3 at 25°C. It is a hygroscopic, strong acid of tart flavor. It is used as an acidulant in fruit drinks and carbonated beverages at 0.25 to 0.40 percent, in cheese at 3 to 4 percent, and in jellies. It is used as an antioxidant in instant potatoes, wheat chips, and potato sticks, where it prevents spoilage by trapping the metal ions. It is used in combination with antioxidants in the processing of fresh frozen fruits to prevent discoloration.

**Citronellal**—A flavoring agent that is a liquid, faintly yellow with an intense odor resembling lemon, citronella, and rose. It is soluble in alcohol and most fixed oils, slightly soluble in mineral oil and propylene glycol, and insoluble in water and glycerin. It is obtained by chemical synthesis; the aldehyde may be obtained from natural oils, such as citronella oil. It is also termed 3,7-dimethyl-6-octen-1-A1.

**Citronellyl Propionate**—A synthetic flavoring agent that is a moderately stable, colorless liquid of light rose-fruity odor. It is practically insoluble in water but is miscible with alcohol. It is stored in glass or tin containers. It has application in baked goods, candy, beverages, and ice cream at 3 to 19 parts per million.

**Citrus Oil**—A flavorant obtained by pressing the oil from the rind of citrus fruits. It is largely composed of terpenes and sesquiterpenes plus the flavor-imparting oxygenated components. It is partly water soluble, not stable, and is used in beverages.

**Clarified Butter**—*See Butter, Clarified*.

**Clarifying Agents**—*See Processing Aids*.

**Clear Flour**—The portion of straight flour (all the flour that can be milled from a wheat blend) that remains after the removal of the patent streams. Clear flours from hard wheat are generally high in ash, dark in color, and high in protein. It is used to increase the strength of flour and is used in rye, dark bread, and pastries.

**Clouding Agents**—*See Processing Aids*.

**Clove**—A spice that is the unripened bud from the clove tree *Eugenia caryophyllata* Thumb. It is very pungent and is used in the whole form in fruit punches, relishes, marinades, and sauces. In the ground form, it is used in cakes, cookies, and meat sauces.

**Coarse-Ground Wheat**—*See Crushed Wheat*.

**Cochineal**—A red colorant extracted from the dried bodies of the female insect *Coccus cacti*. The coloring is carminic acid in which the water-soluble extract is cochineal. It precipitates at pH 3, has good stability at pH 4, and excellent stability at pH 5 to 8. It has low tinctorial strength and has excellent stability to heat and light. It is also stable in retorted protein systems where other food dyes are unstable. It is used in foods requiring red coloring.

**Cocoa Butter**—The fat obtained by pressing chocolate liquor, obtained from roasted cocoa nibs, to yield cocoa butter and presscake. It has a melting point of approximately 33°C but is a hard, brittle solid at room temperature. It is used in the manufacture of coatings for candies, the coatings consisting mainly of mixtures of roasted cocoa nibs, sugar, and cocoa butter. It is also used in confections. It is also termed cacao butter.

**Cocoa Liquor**—The liquor obtained by the grinding of cocoa nibs from the cocoa bean. The liquor converts into cocoa powder and cocoa butter as end products. It is a primary ingredient in chocolate manufacture. It is also termed chocolate liquor.

**Cocoa Powder**—The powder produced by the grinding, pulverizing, and air classification of the cocoa presscake, which is obtained by compressing the cocoa liquor, obtained from cocoa nibs, into a presscake and cocoa butter. There are two main types of powder—alkalized and natural. The alkalized (Dutch processed) has a pH range of 6.5 to 8.1, a red-brown shade which tends to develop red-brown end products, and a mild flavor. It is used in beverages, retail cocoa powder, puddings, and ice cream. The natural has a pH range of 5.2 to 5.9 and a yellow-orange color with a tendency to produce light brown end products. It is used in the baking industry to impart color and flavor and also used in candy, syrups, and toppings.

**Coconut**—The nut obtained from the coconut palm. It provides a source of coconut meat and coconut oil.

**Coconut, Desiccated**—The dried coconut meat whose reduced moisture content increases its stability. It is available in various shapes and sizes. It is used to impart flavor in desserts, baked goods, and candies.

**Coconut Oil**—The oil obtained from the kernel of the nuts of the coconut palm. It has a sharp melting character (narrow plastic range) in that it changes abruptly from a hard, brittle solid to a clear oil with a temperature change of a few degrees, and the transition occurs at room temperature range. It melts at 25°C and is more completely solid than butter at 10°C. These properties make it suited for the preparation of shortenings where brittleness and a large change in consistency with a small temperature change are undesirable. Partially hydrogenated coconut oil has hydrogen added to part of the unsaturated carbon bonds to provide a more solid consistency. It is used in confections, baked goods, and margarine.

**Collagen**—A protein that is the principal constituent of connective tissue and bones of vertebrates; it can be converted to gelatin and glue by boiling in water.

**Colors and Coloring Adjuncts**—Substances used to impart, preserve, or enhance the color or shading of a food, including color stabilizers, color fixatives, color-retention agents, etc. Legally, they are usually designated artificial (synthetic) or natural, which indicates that they are, respectively, synthetically manufactured or obtained from natural sources. Synthetic color additives "certified" by the Food and Drug Administration are designated FD&C (Food, Drug, and Cosmetic). Those acceptable food colors not designated "certified" are designated "approved" and consist of natural organic and synthetic inorganic colorants used in certain applications.

**Confectionary Fat**—A fat that is hard at room temperature and soft at body temperature, such as hydrogenated coconut oil or cocoa butter.

**Copper**—A metal necessary for the maintenance of normal erythropoiesis and the prevention of iron deficiency anemia, iron being essential in hemoglobin synthesis.

**Copper Gluconate**—A light blue powder used as a dietary supplement.

**Copper Sulfate**—A nutrient supplement and processing aid most often used in the pentahydrate form. This form occurs as large, deep blue or ultramarine, triclinic crystals, as blue granules, or as a light blue powder. The ingredient is prepared by the reaction of sulfuric acid with cupric oxide or with copper metal. Copper sulfate may be used in infant formula. It is also termed cupric sulfate.

**Coriander**—A spice that is the dried, ripe fruit of *Coriandrum sativum* L. It has a pleasing, aromatic taste. It is used in sausage, variety meats, and curry powder in the ground form, and in pickles, baked goods, and stuffing in the whole form.

**Corn**—The maize grain, which is the source of various ingredients. It is used in the kernel form for food; it is dry milled into flour, grits, and meal, and it is wet milled into starches, dextrins, dextrose, and other byproducts. The kernel consists of four basic parts which are the starch section, corn germ, gluten, and hull. The starch section comprises approximately 61 percent of the kernel, while the corn germ comprises approximately 4 percent of the kernel. The term *corn* refers to other cereal crops in different areas of the world.

**Corn Bran**—A dry-milled product of high fiber content obtained from corn. It can be used to increase the fiber content of breads, cookies, and cereals, and to thicken gravies and soups.

**Corn Flour**—A finely ground flour made from milling and shifting maize or obtained as a byproduct of cornmeal. It is used as pancake flour.

**Corn Gluten**—A nutrient supplement which is the principal protein component of corn endosperm. It consists mainly of zein and glutelin. Corn gluten is a byproduct of the wet milling of corn for starch. The gluten fraction is washed to remove residual water soluble proteins. Corn gluten is also produced as a byproduct during the conversion of the starch in whole or various fractions of dry milled corn syrups. The ingredient is used in food with no limitation other than current good manufacturing practice. It is also termed corn gluten meal.

**Cornmeal**—A ground corn of specified mesh profile that is made from white or yellow maize. It is used in cornbread mix.

**Corn Oil**—The oil obtained from the germ of the maize plant. The unsaturated fatty acids linoleic and oleic make up 80 to 85 percent of the total fatty acids. The tocopherols prevent the oil from oxidizing rapidly. It has a low melting point of –18 to –10°C. It is used in mayonnaise, margarine, salad oil, and bakery products.

**Corn Silk and Corn Silk Extract**—Flavor agents used in baked goods and baking mixes (30 ppm), nonalcoholic beverages (20 ppm), frozen dairy desserts (10 ppm), soft candy (20 ppm), and all other food categories (4 ppm). Corn silk is the fresh styles and stigmas of *Zea mays* L. Collected when the corn is in milk. The filaments are extracted with dilute ethanol to produce corn silk extract. The extract may be concentrated at a temperature not exceeding 60°C.

**Cornstarch**—The starch made from the endosperm of corn, containing amylose and amylopectin starch molecules. When starch is heated in water it forms a viscous, opaque paste. The paste forms semisolid gels upon cooling and has the ability to form strong adhesive films when spread and dried. Cornstarch is not freeze-thaw stable and is used widely except when clarity or the lack of gel formation is desired. It exists as fine or coarse powders. The coarse starch is sometimes termed pearl starch. It is used in sauces, gravies, puddings, pie fillings, and salad dressings. The typical usage level is 1 to 5 percent. It is also termed maize starch and common cornstarch, regular cornstarch, or unmodified cornstarch.

**Cornstarch, Acid-Modified**—A starch produced by treating suspended cornstarch in water with dilute mineral acid at high temperatures for varying time periods. This is followed by neutralization with sodium carbonate upon obtainment of the desired viscosity.

This produces starches that have decreased viscosity when warm but still form gels when cooled. It is used in the manufacture of starch-based gum candies. It is also termed thin-boiling starch. Esters and ethers can be formed in which only one end of the addition molecule is attached to the starch molecule. These starches have freeze-thaw stability, shear resistance, and acid resistance, and are used in sauces, gravies, and frozen foods.

**Cornstarch, Oxidized**—*See Oxidized Cornstarch*.

**Corn Sugar**—*See Dextrose*.

**Corn Sugar Vinegar**—The product made by the alcoholic and subsequent acetous fermentation of corn sugars according to federal regulations. It is of amber color and has a minimum of 4 percent acid (expressed as acetic acid). It functions as an acidulant in foods.

**Corn Syrup**—A corn sweetener that is a viscous liquid containing maltose, dextrin, dextrose, and other polysaccharides. It is obtained from the incomplete hydrolysis of cornstarch. It is classified according to the degree of conversion which is expressed as the dextrose equivalent (DE), which is the measure of sweetness of the corn syrup as compared to that of a sucrose syrup. Generally the greater the degree of conversion, the sweeter the syrup. Corn syrup is used as a replacement for sucrose but is less sweet than sucrose. It can control crystallization in candy making, contribute body in ice cream, and provide pliability in confections. It is also termed glucose syrup.

**Corn Syrup Solids**—The dry form of corn syrup used where it is impractical to use the liquid syrup. *See Corn Syrup*.

**Cracked Wheat**—The wheat prepared by cracking or cutting cleaned wheat, other than durum wheat and red durum wheat, into angular fragments. The proportions of the natural constituents, other than moisture, remain unaltered. The moisture content does not exceed 15 percent.

**Cracker Flour**—Flour that is long patented or straight grades of soft wheat flour, containing 9 to 10.5 percent protein. It is of low absorption and has short mixing requirements.

**Cranberry Extract**—A natural red colorant with good pH stability and fair heat, light, and chemical stability. The anthocyanidin pigments in cranberry are peonidin and cyanidin. The extract has low tinctorial strength and good stability at pH 3 to 4. Because the color is affected by pH, it can only be used in acidic mediums such as beverages.

**Cream**—That portion of milk that is high in milkfat and will rise to the top of undisturbed milk. It is obtained by the separation of the fat fraction of the milk to concentrations ranging from 18 to 40 percent fat. Cream is labeled according to the fat content: heavy whipping cream has a minimum of 36 percent fat; light whipping cream has 30 to 36 percent fat; and light, coffee, or table cream has 18 to 30 percent fat. The lower fat creams are usually prepared by blending a high-fat cream with milk. Cream is used in ice cream mix, whipped toppings, and sauces.

**Cream of Tartar**—The acid potassium salt of tartaric acid occurring as crystals or powder. It is relatively poorly soluble having a solubility in 100 ml of water at 0.8 g at 25°C and 6.1 g at 100°C. A 1 percent solution at 30°C has a pH of 3.4. Chemical names are potassium acid tartrate, potassium hydrogen tartrate, and potassium bitartrate. It functions to complex with heavy metal ions and to regulate pH; it can have a gentle laxative action if given at adequate levels. The acidulant is used in chemical leavening to release carbon dioxide which produces the loaf volume. It has limited reactivity in the cold so when used in reduced-temperature batters it has little gas evolution during the initial mixing. At room temperatures, it has a relatively fast reaction rate. It functions as a taste regulator in sugar icing and in controlled crystallization of toffees and fondants by the regulated inversion of sucrose. It is used in baked goods, crackers, candy, and puddings.

**Cresyl Acetate**—A flavoring agent that is a clear and colorless liquid with a strong, flowery odor. It is soluble in most fixed oils and propylene glycol, moderately soluble in mineral oil, and insoluble in glycerin. It is obtained by chemical synthesis. It is also termed p-toyl acetate.

**Crown Gum**—A product of vegetable origin used as a masticatory substance in chewing gum base.

**Crushed Wheat**—The wheat prepared by crushing cleaned wheat, other than durum wheat and red durum wheat. The proportions of the natural constituents, other than moisture, remain unaltered. The moisture content does not exceed 15 percent. It is also termed coarse-ground wheat.

**Cumin**—A spice that is the dried, ripe fruit of *Cuminum cyminum* L. It is usually obtained in the ground form. It has a warm, pleasant, balsamic flavor. It is used in cheese, soups, relishes, and meats.

**Cuminic Aldehyde**—A flavoring agent that is a liquid, colorless to yellow in appearance, with a strong pungent odor resembling cumin oil. It is insoluble in water and soluble in alcohol and ether. It is obtained from cumin oil. It is also termed p-cuminic aldehyde, cumaldehyde, and cuminal.

**Curdlan**—A polymer produced by fermentation from the bacteria *Alcaligenes faecalis* var. *myxogenes*. It forms a gel when heated in water, initially requiring to be dispersed and suspended in water since it is insoluble until heated. Above 80°C a high-set gel that is irreversible and stable to freezing is formed. A low-set gel is obtained by heating to 55 to 60°C and then cooling below 40°C; this gel is thermoreversible. It is capable of gelling over pH range of 2 to 10. The heat gelling properties allow improved water retention in sausage and ham, improved consistency in Oriental-style noodles, and provides viscosity and texture.

**Curry Powder**—A blend of spices used as seasoning in curries, sauces, and meats. Typical spices in the blend include coriander, ginger, nutmeg, clove, cinnamon, red pepper, and onion salt.

**Cyanocobalamin**—Vitamin $B_{12}$, a water-soluble vitamin required for the normal development of red blood cells. Its deficiency causes pernicious anemia. It is stable in neutral conditions and is more stable for storage than for processing conditions. It is found in meat, fish, and milk.

**Cyclohexyl Acetate**—A synthetic flavoring agent that is a stable, colorless liquid of fruity odor. It is stored in glass, tin, or resin-lined containers. It is used for flavors such as apple, banana, blackberry, and raspberry with applications in beverages, ice cream, candy, and baked goods at 15 to 110 parts per million.

**Cyclohexyl Butyrate**—A synthetic flavoring agent that is a stable, colorless liquid of fruity odor. It should be stored in glass, tin, or resin-lined containers. It is used in pineapple, apricot, banana, and berry flavor with applications in beverages, ice cream, and candy at 4 to 9 parts per million.

**Cyclohexyl Cinnamate**—A synthetic flavoring agent that is a stable, colorless to light yellow liquid of fruity odor. It is stored in glass or tin containers. It is used in peach and cherry flavors with applications in ice cream, candy, and baked goods at 5 to 20 parts per million.

**Cyclohexyl Formate**—A synthetic flavoring agent that is a moderately stable, colorless liquid of fruity odor. It should be stored in glass or tin containers. It is used in apple or plum flavors and gives a lift to fruity flavors. It has application in beverages, candy, ice cream, and baked goods at 3 to 11 parts per million.

**Cyclohexyl Propionate**—A synthetic flavoring agent that is a stable, colorless liquid of fruity odor. It is stored in glass or tin containers. It is used in fruit flavors such as pineapple with applications in beverages, candy, ice cream, and baked goods at approximately 3 parts per million.

**Cydonia Seed**—*See Quince Seed*.

**Cysteine**—A nonessential amino acid that functions as a nutrient and dietary supplement. It is used in foods to prevent oxygen from destroying vitamin C and is used in doughs to reduce mixing time.

**Cystine**—A nonessential amino acid that acts as a nutrient and dietary supplement. It is very slightly soluble in water and in alcohol. It improves the biological quality of the total protein in foods containing naturally occurring intact protein.

# D

**d-Limonene**—A flavoring agent that is a liquid, colorless with a pleasant odor resembling mild citrus. It is miscible in alcohol, most fixed oils, and mineral oil; soluble in glycerin; and insoluble in water and propylene glycol. It is obtained from citrus oil. It is also termed d-p-mentha-1,8,diene and cinene.

**Danish Agar**—*See Furcelleran*.

**1-Decanol, Natural**—A flavoring agent that is a liquid, colorless, with a flowery odor similar to orange blossoms. It is insoluble in water and glycerin, and soluble in alcohol, ether, and mineral oil.

**Defoaming Agents**—*See Surface-Active Agents*.

**Dehydroacetic Acid**—(DHA) A preservative that is a crystalline powder with a solubility of less than 0.1 g in 100 g water at 25°C. It can undergo a variety of chemical reactions which give it utility in many applications. It is used at 0.01 to 0.5 percent for microbiological growth inhibition in various foods. It is used for cut or peeled squash, with no more than 65 parts per million remaining in or on the prepared squash.

**Dextrin**—A partially hydrolyzed starch formed from the treatment of starch by dry heat, acid, or enzymes. It can be formed from amylose- and amylopectin-type starches. They are white or yellow (canary) in color. As compared to unmodified starch, dextrins have increased water solubility, viscosity stability, and reduced paste viscosity. Uses include dough improvement and binding.

**Dextrose**—A corn sweetener that is commercially made from starch by the action of heat and acids or enzymes, resulting in the complete hydrolysis of the cornstarch. There are two types of refined dextrose commercially available: dextrose hydrate, which contains 9 percent by weight water of crystallization and is the most often used, and anhydrous dextrose, which contains less than 0.5 percent water. Dextrose is a reducing sugar and produces a high-temperature browning effect in baked goods. It is used in ice cream, bakery products, and confections. It is also termed glucose and corn sugar.

**Diacetyl Tartaric Acid Esters of Mono- and Diglycerides**—A hydrophilic emulsifier used in oil-in-water emulsions. The connecting of glycerol with tartaric acid prior to esterification of the other part of the glycerol increases the hydrophilicity of the emulsifier. It functions as a dough conditioner in freestanding breads and rolls to strengthen the gluten which improves crumb softness, crust, and increased volume. It is used in coffee whiteners for dispersion. It is used in chocolate couverture to adjust the consistency, viscosity, and adhesion ability. In reduced-calorie breads, it reduces the quantity of shortening required and maintains volume. It is also termed acetylated tartaric acid monoglyceride, and acetyl tartrate mono- and diglyceride.

**Dibasic Calcium Phosphate, Anhydrous**—*See Dicalcium Phosphate, Anhydrous.*

**Dibasic Calcium Phosphate, Dihydrate**—*See Dicalcium Phosphate, Dihydrate.*

**Dicalcium Phosphate, Anhydrous**—A mineral supplement and dough conditioner. It contains approximately 29 percent calcium. It is practically insoluble in water, with a solubility of 0.02 g per 100 ml water at 25°C. It is also termed calcium phosphate, dibasic anhydrous and dibasic calcium phosphate, anhydrous. It is used as a mineral supplement in prepared breakfast cereals, enriched flour, and noodle products.

**Dicalcium Phosphate, Dihydrate**—A source of calcium and phosphorus that also functions as a dough conditioner and bleaching agent. It functions as a dough conditioner in bakery products, as a bleaching agent in flour, as a source of calcium and phosphorus in cereal products, and as a source of calcium for alginate gels. It contains approximately 23 percent calcium. It is practically insoluble in water. It is also termed dibasic calcium phosphate, dihydrate and calcium phosphate dibasic, hydrous. It is used in dessert gels, baked goods, cereals, and breakfast cereals.

**Diacetyl**—A flavoring agent that is a clear yellow to yellowish green liquid with a strong pungent odor. It is also known as 2,3-butanedione and is chemically synthesized from methyl ethyl ketone. It is miscible in water, glycerin, alcohol, and ether, and in very dilute water solution it has a typical buttery odor and flavor.

**Diethyl Sebacate**—A flavoring agent that is a liquid, colorless to pale yellow in appearance with a slight odor. It is insoluble in water and miscible in alcohol, ether, and other organic solvents. It is obtained by chemical synthesis. It is also termed ethyl sebacate.

**Diglyceride**—A lipophilic emulsifier prepared by direct esterification of two fatty acids with glycerol, or by interesterification between glycerol and other triglycerides. It often occurs as a blend with monoglycerides. It is widely used in numerous foods such as ice cream, puddings, margarine, doughs, shortenings, peanut butter, and coffee whiteners. It has numerous functions including the provision of dough conditioning, prevention of fat separation, and the provision of emulsion stability and dispersibility.

**Dilauryl Thiodipropionate**—(DLTDP) An antioxidant that exists as white crystalline flakes of sweetish ester-like odor. It is insoluble in water but soluble in inorganic solvents. It is used in fats and oils to prevent rancidity. It is used in peanut oil at a maximum usage level of 200 mg/kg.

**Dill and Its Derivatives**—A flavoring agent that is the herb and seeds from *Anethum graveolens* L., dill (Indian), and the herb and seeds from *Anethum sowa*, D.C. Its derivatives include essential oils, oleoresins, and natural extractives obtained from these sources of dill.

**Dill Seed**—A spice that is the dried, ripe fruit of the plant *Anethum graveolens* L. It is extremely pungent and slightly dominant. It is used in dips, spreads, sauces, and meats.

**Dill Weed**—A spice made from the leaf of the dill plant. While dill seed has a camphorous, slightly bitter taste and fragrance, the weed has a delicate bouquet which enhances rather than dominates. It is used in meats and sauces.

**Dimethylpolysiloxane**—An antifoaming agent used in fats and oils. It prevents foaming and spattering when oils are heated and prevents foam formation during the manufacture of wine, refined sugar, gelatin, and chewing gum. It is also termed methyl polysilicone and methyl silicone.

**Dioctyl Sodium Sulfosuccinate**—A wetting and emulsifying agent that is slowly soluble in water, having a solubility of 1 g in 70 ml of water. It functions as a wetting agent in fumaric acid–containing powdered fruit drinks to help the acid dissolve in water. It is used as a stabilizing agent on gums at not more than 0.5 percent by weight of the gum. It is used as a flavor potentiator in canned milk where it improves and maintains the flavor of the sterilized milk during storage. It also functions as a processing aid in the manufacture of unrefined sugar. It is also termed sodium dioctylsulfosuccinate.

**Dipotassium Monohydrogen Orthophosphate**—*See Dipotassium Phosphate.*

**Dipotassium Monophosphate**—*See Dipotassium Phosphate.*

**Dipotassium Phosphate**—The dipotassium salt of phosphoric acid which functions as a stabilizing salt, buffer, and sequestrant. It is mildly alkaline with a pH of 9 and is soluble in water with a solubility of 170 g per 100 ml of water at 25°C. It improves the colloidal solubility of proteins. It acts as a buffer against variation in pH. For example, it is used in coffee whiteners as a buffer against pH variation in hot coffee and to prevent feathering. It also functions as an emulsifier in specified cheeses and as a buffering agent for processed foods. It is also termed dipotassium monohydrogen orthophosphate, potassium phosphate dibasic, and dipotassium monophosphate.

**Disodium Calcium EDTA**—A sequestrant and chelating agent whose complete name is disodium calcium ethylenediamine tetraacetate. It is a nonhygroscopic powder that is colorless, odorless, and tasteless at recommended use levels. A 1 percent solution has a pH of 6.5 to 7.5. It is used to control the reaction of trace metals with some organic and inorganic components in food; to prevent deterioration of color, texture, and development of precipitates; and to prevent oxidation. Its function is comparable to that of disodium dihydrogen EDTA. It is also termed calcium disodium EDTA. *See EDTA.*

**Disodium Dihydrogen EDTA**—A sequestrant and chelating agent whose complete name is disodium ethylenediamine tetraacetate. It is a nonhygroscopic powder that is colorless, odorless, and tasteless at recommended use levels. A 1 percent solution has a pH of 4.3 to 4.7. It is used to control the reaction of trace metals to include calcium and magnesium with other organic and inorganic components in food to prevent deterioration of color, texture, and development of precipitates and to prevent oxidation. Its function is comparable to that of disodium calcium EDTA. *See EDTA.*

**Disodium Dihydrogen Pyrophosphate**—*See Sodium Acid Pyrophosphate.*

**Disodium Diphosphate**—*See Sodium Acid Pyrophosphate.*

**Disodium Guanylate**—A flavor enhancer which is a crystalline powder, colorless or white, and has characteristic taste. It is soluble in water, sparingly soluble in alcohol, and practically insoluble in ether. It is obtained by chemical synthesis. It is also termed sodium 5'-guanylate and disodium guanosine-5'-monophosphate.

**Disodium 5'-Inosinate**—A flavor enhancer which performs as a disodium guanylate does, but only when present at approximately twice the level. *See Disodium Guanylate.*

**Disodium Monohydrogen Orthophosphate**—*See Disodium Phosphate.*

**Disodium Monohydrogen Orthophosphate Dihydrate**—*See Disodium Phosphate.*

**Disodium Monophosphate**—*See Disodium Phosphate.*

**Disodium Phosphate**—The disodium salt of phosphoric acid which functions as a protein stabilizer, buffer, dispersant, and coagulation accelerator. It is mildly alkaline with a 1 percent solution having a pH of 9.2. It is moderately soluble in water with a solubility of 12 g in 100 ml at 25°C. It is used in farina and macaroni to shorten the cooking time by making the particles swell faster and cook more thoroughly. In evaporated milk it acts as a buffer and prevents gelation, also acting as a buffer in coffee whiteners. It is an accelerator of the setting time in instant puddings. In cream sauce and whipped products it functions as a dispersant by producing a swelling of protein. It is also termed disodium monohydrogen orthophosphate; disodium monohydrogen orthophosphate dihydrate; sodium phosphate, dibasic; sodium phosphate, dibasic dihydrate; and disodium monophosphate.

**Disodium Phosphate, Duohydrate**—An emulsifier, buffer, and mineral supplement. It is moderately soluble in water with a solubility of 15 g in 100 ml of water at 25°C. A 1 percent solution has a pH of 9.1. It is used in processed cheese for uniform texture and smoothness. It is also termed disodium phosphate, dihydrate and sodium phosphate dibasic, dihydrate.

**Disodium Tartrate**—*See Sodium Tartrate.*

**Dispersants—*See Surface-Active Agents*.**

**Distilled Monoglyceride—**An emulsifier containing a minimum of 90 percent monoglyceride derived from edible fat and glycerine. It is an active monoglyceride produced by distillation to obtain the monoglyceride fraction, which is the part that functions as an emulsifier or food quality improver. Commercially termed monoglycerides also contain diglycerides, triglycerides, and so on. It is used in margarine, peanut butter, shortenings, bakery goods, and whipped desserts to improve texture and consistency. Typical usage levels are 0.1 to 1.0 percent.

**Distilled Vinegar—*See Vinegar, Distilled*.**

**Dodecyl Gallate—**An antioxidant used in cream cheese, instant mashed potatoes, margarine, fats, and oils.

**Dough Conditioner—**A blend of minerals used in baked goods. It is usually contained within yeast foods as a blend of calcium salts, sulfates, and phosphates which toughen the gluten. Usage of hard water generally results in better breads so the minerals serve to minimize the effect of variables in water conditions. It is also termed yeast food.

**Dough Strengtheners—**Substances used to modify starch and gluten, thereby producing a more stable dough.

**Dried Buttermilk—*See Buttermilk, Dried*.**

**Dried Milk—*See Whole Milk Solids*.**

**Dried Skim Milk—*See Milk Solids–Not-Fat*.**

**Dry Ice—*See Carbon Dioxide*.**

**Dry Whole Milk—*See Milk Powder*.**

**Durum Flour—**The fine powder obtained from durum wheat, which is fine enough to pass through a number 100 U.S. sieve. It is used principally in macaroni and spaghetti products because it provides the desired texture and consistency. ***See Durum Wheat*.**

**Durum Granular—**The product obtained from durum wheat by grinding to obtain semolina to which flour is added so that 7 to 20 percent passes through a number 100 U.S. sieve. It is used in macaroni and spaghetti. ***See Durum Wheat*.**

**Durum Semolina—*See Semolina*.**

**Durum Wheat**—The wheat obtained from the durum wheat kernel. It differs from other hard wheats in that the starch swelling capacity is greater and the gluten has different characteristics which result in tough, elastic doughs. As compared to hard wheat dough, it can be extruded through a small hole at lower pressure and in breads results in lower loaf volume. It is used almost exclusively in macaroni and spaghetti products because it is easily processed to produce a smooth, mechanically strong product of desired color which when cooked will maintain its shape and be of firm consistency. Products derived from the wheat include durum flour, durum granular, and durum semolina.

# E

**EDTA**—The abbreviation for ethylenediaminetetraacetate, a sequestrant and chelating agent that functions in water but not fats and oils. It is used to control the reaction of trace metals with some organic and inorganic components to prevent deterioration of color, texture, and development of precipitates, as well as to prevent oxidation which results in rancidity. The reactive sites of the metal ions are blocked, which prevents their normal reactions. The most common interfering metal ions in food products are iron and copper. It can be used in combination with the antioxidants BHT and propyl gallate. It is used in margarine, mayonnaise, and spreads to prevent the vegetable oil from going rancid. It is used in canned corn prior to retorting to prevent discoloration caused by trace quantities of copper, iron, and chromium. It also inhibits copper-catalyzed oxidation of ascorbic acid. It occurs as disodium calcium EDTA and disodium dihydrogen EDTA. Its use is approved in specified foods, with an average usage level being in the range of 100 to 300 parts per million.

**Egg**—The hard-shelled reproductive body of poultry. The shell is largely composed of calcium carbonate, and represents approximately 11 percent of its total weight. Inside the shell are the shell membranes, which are principally protein. The yolk, which represents approximately 31 percent of the egg's weight, contains protein, fat, and all the known vitamins except vitamin C. Most of the egg's calories come from the yolk. The egg white is protein and represents approximately 58 percent of the weight. The white does not appear white in color until beaten or cooked. There is a thick and thin white, which consists mainly of ovalbumin, conalbumin, ovoglobulin, ovomucoid, and ovomucin. Eggs are used whole, as egg white, as yolk, or any combination thereof. They are used for coagulation, foam formation, emulsification, nutrition, flavor, and color.

**Egg Albumen**—The protein fraction of egg, which is also termed egg white. It represents approximately 65 percent of the edible egg and is composed of approximately 87 percent water, 11 percent protein, and 1 percent carbohydrate. It provides a source of protein and provides foam upon whipping. It is used in meringues, cakes, and desserts.

**Egg White**—*See Egg Albumen*.

**Egg Yolk**—The yellow portion of the egg, representing approximately 35 percent of the edible egg. It is composed of approximately 49 percent water, 16 percent protein, 32 percent fat, and trace carbohydrate. It is used as an emulsifier in mayonnaise, salad dressing, and cream puffs. It is also used as a source of color.

**Emulsifiers and Emulsifier Salts**—Substances which reduce the surface tension between two immiscible phases at their interface, allowing them to become miscible. The interface can be between two liquids, a liquid and a gas, or a liquid and a solid. Most emulsions involve water and oil or fat as the two immiscible phases, one being dispersed as finite globules in the other. The liquid as globules is referred to as the dispersed or internal phase, while the medium in which they are suspended as the continuous or external phase. There are two types of oil/water emulsions depending on the composition of the phases. In an oil-in-water emulsion such as milk and mayonnaise, the water is the external phase and the oil is the internal phase. In a water-in-oil emulsion such as butter, the oil is the external phase and the water is the internal phase. Emulsifiers have the following major functions:
- Complexing—Reaction with starch in bakery products which retards the crystallization of the starch, thus retarding the firming of the crumb which is associated with staling.
- Dispersing—The reduction of interfacial tension which creates an intimate mixture of two liquids that normally are immiscible, an example being oil-in-water emulsions such as salad dressing.
- Crystallization control—Control of crystallization in sugar and fat systems, i.e., chocolate, where it allows for brighter initial gloss and prevention of solidified fat on the surface.
- Wetting—Allows the surface to be more attracted to water, such as powders, i.e., coffee whitener, in which the addition of surfactant aids the dispersion of the powder in the liquid without lumping on the surface.
- Lubricating—Functions as a lubricant, such as in caramels, by reducing their tendency to stick to cutting knives, wrappers, and teeth.

**Enriched Bleached Flour**—Flour that has been whitened by removal of the yellow pigments and fortified with vitamins and minerals. The added vitamins are thiamin, riboflavin, niacin, or niacinamide, and may include vitamin D. The minerals are iron and may include calcium. It is used in baked goods.

**Entire Wheat Flour**—*See Whole Wheat Flour.*

**Epsom Salt**—*See Magnesium Sulfate.*

**Ergosterol**—A steroid alcohol that when irradiated with ultraviolet light yields calciferol (Vitamin $D_2$). Irradiated ergosterol is added to milk for vitamin D fortification.

**Erythorbic Acid**—A food preservative that is a strong reducing agent (oxygen accepting) which functions similarly to antioxidants. In the dry crystalline state it is nonreactive, but in water solutions it reacts readily with atmospheric oxygen and other oxidizing agents, making it valuable as an antioxidant. During preparation, dissolving and mixing should incorporate a minimum amount of air, and storage should be at cool temperatures. It has a solubility of 43 g per 100 ml of water at 25°C. One part erythorbic acid is equivalent to 1 part ascorbic acid and equivalent to 1 part sodium erythorbate. It is used to control oxidative color and flavor deterioration in fruits at 150 to 200 parts per million. It is used in meat curing to speed and control the nitrite curing reaction and prolong the color of cured meat at levels of 0.05 percent.

**Erythritol**—A sweetener (polyol) manufactured by fermentation of glucose, the glucose-rich substrate being obtained by the enzymatic hydrolysis of starch. It is 60 to 70 percent as sweet as sugar, has excellent heat and acid stability, a high digestive tolerance, and a caloric value of 0.2 Kcal/gram. It is the only polyol produced by fermentation. It can be used as a sugar replacement in confectioneries, beverages, and desserts.

**Erythrosine**—*See FD&C Red #3.*

**Ethoxylated Mono- and Diglycerides**—An emulsifier prepared by the glycerolysis of edible vegetable fats and reacting with ethylene oxide. It is hydrophilic, being soluble in water and partially soluble in oil. It contributes to freeze-thaw stability and overrun in whipped toppings. It functions as a dough conditioner/emulsifier in baked goods and as an emulsifier in coffee whiteners, icings, and frozen desserts. Typical usage levels are 0.20 to 0.45 percent. It is also termed polyglycerate 60 and polyoxyethylene (20) mono- and diglycerides of fatty acids.

**Ethoxyquin**—An antioxidant used in the preservation of color in chili powder, ground chili, and paprika.

**Ethyl Acrylate**—A flavoring agent that is a clear, colorless liquid. Its odor is fruity, harsh, penetrating, and lachrymatous (causes tears). It is sparingly soluble in water and miscible in alcohol and ether, and is obtained by chemical synthesis.

**2-Ethylbutyric Acid**—A flavoring agent that is a clear liquid, colorless, with a rancid odor. It is miscible in alcohol and ether, sparingly soluble in water, and is obtained by chemical synthesis.

**Ethyl Cellulose**—Used as a binder and filler in dry vitamin preparations, as a component of protective coatings for vitamin and mineral tablets, and as a fixative in flavoring compounds. It is a cellulose ether containing ethyoxy groups attached by an ether linkage and containing an anhydrous basis of not more than 2.6 ethoxy groups per anhydroglucose unit.

**Ethyl Crotonate**—A synthetic flavoring agent that is a moderately stable, colorless to light yellow liquid of sharp winey note. It should be stored in glass, tin, or resin-lined containers. It is used in fruit flavors for application in baked goods, beverages, and candy at 2 to 7 parts per million.

**Ethylenediaminetetraacetate**—*See EDTA*.

**Ethylene Oxide Polymer**—Foam stabilizer in fermented malt beverages which is the polymer of ethylene oxide. It is used at a level not to exceed 300 parts per million by weight of the fermented malt beverage. The label of the additive bears directions for use to ensure compliance with the legal limit.

**Ethyl Formate**—A flavoring agent that occurs naturally in some plant oils, fruits, and juices but does not occur naturally in the animal kingdom. It is used in food at a maximum level, as served, of 0.05 percent in baked goods; 0.04 percent in chewing gum, hard candy, and soft candy; 0.02 percent in frozen dairy desserts; 0.03 percent in gelatins, puddings, and fillings; and 0.01 percent in all other food categories. It is an ester of formic acid and is prepared by esterification of formic acid with ethyl alcohol or by distillation of ethyl acetate and formic acid in the presence of concentrated sulfuric acid.

**Ethyl-2,4-Hexadieonate—*See Ethyl Sorbate*.**

**Ethyl Isobutyrate**—A synthetic flavoring agent that is a stable, colorless liquid of dry, fruity odor. It should be stored in tin, glass, or resin-lined containers. It is used to give fruity effects to flavors for applications in candy, baked goods, and beverages at 10 to 100 parts per million.

**Ethyl Lactate**—A solvent manufactured from L(+) lactic acid which is miscible in water and most organic solvents and is cleared for use as a flavoring agent. It is a naturally occurring constituent of California and Spanish sherries. It is a clear, colorless, nontoxic liquid of low volatility, having a pH of 7 to 7.5. It is used as a food and beverage flavoring agent.

**Ethyl Maltol**—A flavoring agent that is a white, crystalline powder. It has a unique odor and a sweet taste that resembles fruit. The melting point is 90°C. It is sparingly soluble in water and propylene glycol and soluble in alcohol and chloroform. It is obtained by chemical synthesis.

**Ethyl-Methyl-Phenyl-Glycidate**—A synthetic flavoring agent that is a glycidic acid ester. It is a colorless to pale yellow liquid with a strong fruit odor suggestive of strawberries. It is unstable to alkali and moderately stable to weak organic acids. It should be stored in glass, tin, or aluminum containers. It is soluble in fixed oils and in propylene glycol. It is used in flavors for strawberry note and has application in candy, beverages, and ice cream at 6 to 20 parts per million. It is also termed aldehyde C-16.

**Ethyl Nonanoate**—A synthetic flavoring agent that is a stable, colorless liquid of fruit cognac odor. It is practically insoluble in water and is miscible with alcohol and propylene glycol. It should be stored in glass or tin containers. It is used in flavors such as apple, pear, and cognac with applications in beverages, ice cream, candy, and alcohol beverages at 4 to 20 parts per million.

**Ethyl Oxyhydrate—*See Rum Ether*.**

**Ethyl Paraben—*See Parabens*.**

**Ethyl Propionate**—A flavoring agent that is a transparent liquid, colorless, with an odor resembling rum. It is miscible in alcohol and propylene glycol, soluble in fixed oils, mineral oil, and alcohol, and sparingly soluble in water. It is obtained by chemical synthesis.

**Ethyl Sorbate**—A synthetic flavoring agent that is a moderately stable, light yellow liquid of fruity odor. It should be stored in glass or tin containers. It is used in flavors such as pineapple, papaya, and passion fruit with applications in ice cream, beverages, candy, and baked goods at 6 to 18 parts per million. It is also termed ethyl-2,4-hexadienoate.

**Ethyl Vanillin**—A flavoring agent that is a synthetic vanilla flavor with approximately three and one-half times the flavoring power of vanillin. It has a solubility of 1 g in 100 ml of water at 50°C. It is used in ice cream, beverages, and baked goods.

**Eugenol**—A flavoring obtained from clove oil and also found in carnation and cinnamon leaves. It is a stable, light yellow-green liquid of clove odor. It is slightly soluble in water and miscible in alcohol. It should be stored in glass or tin, avoiding iron containers. It is used in spice oils for application in condiments and meats at 100 to 200 parts per million and in baked goods and candy at approximately 30 parts per million.

**Extract**—An alcohol or alcohol-water solution that contains a flavoring ingredient obtained from a spice or some other ingredient and which is used as a flavorant. It is used in baked goods, beverages, and ice cream.

**Extract of Malted Barley and Corn**—*See Malted Cereal Syrup*.

# F

**Family Flour—***See All-Purpose Flour*.

**Farina**—Wheat, other than durum or red durum wheat, from which the bran and most of the germ has been removed. It is ground so that not more than 3 percent passes through a number 100 U.S. sieve.

**Fast Green FCF—***See FD&C Green #3*.

**Fat**—Water-insoluble material of plant or animal origin, consisting predominantly of glyceryl esters of fatty acids (triglycerides). Fat ordinarily refers to triglycerides that are semisolid at room temperature. Fat in its liquid state is called oil.

**Fatty Acids**—Aliphatic acids that may be saturated or unsaturated, consisting of a mixture of certain monobasic carboxylic acids and their associated fatty acids. Fatty acids plus glycerol result in a fat characterized by the fatty acid components. A fatty acid may be used as a lubricant, a binder, a food processing defoamer, and an emulsifier.

**FD&C Blue #1**—A colorant. It has a solubility in water of 20 g in 100 ml at 25°C. It has good pH stability with only slight fading after one week at pH 3, 7, and 8 but is unstable in alkalis such as 10 percent sodium carbonate and 10 percent ammonium hydroxide. It has good stability in 10 percent sugar systems. It has fair stability to light, fair to poor stability to oxidation, and good stability to heat. It has a greenish-blue hue with excellent tinctorial strength. It is used with other primary colors to produce a variety of shades, for example, in combination with FD&C Yellow #5, it gives green. It has good compatibility with food components. It is used in candies, baked goods, soft drinks, and desserts. The common name is brilliant blue FCF.

**FD&C Blue #2**—A colorant. It has poor pH stability in that after one week at pH 3 and 5 it will appreciably fade, at pH 7 considerably fade, and at pH 8 fade completely. It is the least soluble of all food colors, with a solubility of 1.6 g in 100 ml of water at 25°C. Complete fading occurs in alkalis such as 10 percent sodium carbonate and 10 percent sodium hydroxide, with fading also occurring in 10 percent sugar systems. It has very poor light stability and oxidation stability, and moderate stability to heat; it has a deep blue hue with poor tinctorial strength. It is the only food color that has good resistance to reducing agents, but has very poor compatibility with food components. The major use is in pet food, but it is also used in candies, confections, and baked goods. The common name is indigotine.

**FD&C Green #3**—A colorant. It has good pH stability, showing after one week a slight fade at pH 3, a very slight fade at pH 5 to 7, and slight fade and appreciably bluer color at pH 8. It has excellent solubility in water with a solubility of 20 g in 100 ml at 25°C. It has fair to good stability to light, poor stability to oxidation, and shows no appreciable change in 10 percent sugar systems. It has a bluish-green hue, with excellent tinctorial strength. It has good compatibility with food components and is occasionally used in cereals, soft drinks, beverages, and desserts. The common name is fast green FCF.

**FD&C Red #3**—A colorant. It is not recommended for use below pH 5.0, being insoluble at pH 3 to 5 but being stable at pH 7 and 8. It has a solubility in water of 9 g per 100 ml at 25°C. It has fair stability to oxidation and poor to fair stability to light, while having good stability in 10 percent sugar systems. It has exceptional clarity and brilliance, having a bluish pink hue with very good tinctorial strength. It has poor compatibility with food components and is used in candies and confections as well as cherry dyeing. The common name is erythrosine.

**FD&C Red #40**—A colorant. It has good stability to pH changes from pH 3 to 8, showing no appreciable change. It has excellent solubility in water with a solubility of 22 g per 100 ml at 25°C. It has very good stability to light, fair to poor stability to oxidation, good stability to heat, and shows no appreciable change in stability in 10 percent sugar systems. It has a yellowish-red hue and has a very good tinctorial strength. It has very good compatibility with food components and is used in beverages, desserts, candy, confections, cereals, and ice cream. The common name is Allura® red AC.

**FD&C Yellow #5**—A colorant. It has good stability to changes in pH, showing no appreciable change at pH 3 to 8. It has excellent solubility in water with a solubility of 20 g in 100 ml at 25°C. It has good stability to light and heat, fair stability to oxidation, and shows no appreciable change in 10 percent sugar systems. It has a lemon-yellow hue and has good tinctorial strength. It has moderate compatibility with food components and is used in beverages, baked goods, pet foods, desserts, candy, confections, cereal, and ice cream. The common name is tartrazine.

**FD&C Yellow #6**—A colorant. It has good stability to changes in pH, showing no appreciable change at pH 3 to 8. It has excellent solubility in water with a solubility of 19 g in 100 ml at 25°C. It has moderate stability to light, fair stability to oxidation, good stability to heat, and shows appreciable change in 10 percent sugar systems. It has a reddish-yellow hue and has good tinctorial strength. It has moderate compatibility with food components and is used in beverages, bakery goods, dessert confections, and ice cream. The common name is sunset yellow FCF.

**Fennel**—A spice that is the dried, ripe fruit of the herb *Foeniculum vulgare* Mil. It is a seed with licorice flavor. It is used in meat, fish, and sauces as a seasoning.

**Fenugreek**—The seed, usually in ground form, of the herb *Trigonella foenumgraecum*. It has a maple-like flavor and burnt sugar taste. It is used in curry powder, imitation maple flavor, chutney, and pickles.

**Ferric Ammonium Citrate**—A nutrient and dietary supplement that is a source of iron, containing 17 percent iron.

**Ferric Chloride**—A nutrient and dietary supplement that serves as a source of iron.

**Ferric Citrate**—A nutrient supplement that is prepared from reaction of citric acid with ferric hydroxide. It is a compound of indefinite ratio of citric acid and iron. The ingredient may be used in infant formula. It is also termed iron (III) citrate.

**Ferric Orthophosphate**—An inert white powder that is a source of iron and produces no discoloration or rancidity. It contains approximately 28 percent iron. It is used as a mineral supplement where rancidity is not a problem. It is used in frozen egg substitute, pasta products, and rice products.

**Ferric Oxide**—A nutrient and dietary supplement that is a source of iron.

**Ferric Phosphate**—A nutrient supplement that is an odorless, yellowish white to buff-colored powder and contains from one to four molecules of water of hydration. It is prepared by reaction of sodium phosphate with ferric chloride or ferric citrate. It is also termed iron (II) phosphate.

**Ferric Pyrophosphate**—A nutrient supplement, tan or yellowish white in color, prepared by reacting sodium pyrophosphate with ferric citrate. The ingredient may be used in infant formula. It is also termed iron (III) pyrophosphate.

**Ferric Sulfate**—A nutrient and dietary supplement that is a source of iron.

**Ferrous Ascorbate**—A nutrient supplement, blue-violet in color, containing 16 percent iron. It is a reaction product of ferrous hydroxide and ascorbic acid. It may be used in infant formula.

**Ferrous Carbonate**—A nutrient and dietary supplement that is a source of iron.

**Ferrous Citrate**—A nutrient and dietary supplement that is a source of iron.

**Ferrous Fumarate**—A reddish orange to red-brown powder that is a source of iron. It has high bioavailability and can be used in foods where the red color can be masked. It contains approximately 33 percent iron. It is used as a dietary supplement in breakfast cereals, poultry stuffing, enriched flour, and instant drinks.

**Ferrous Gluconate**—A nutrient and dietary supplement that is a source of iron and a coloring adjunct. It is a yellowish gray to pale greenish yellow powder or granules with a burnt sugar odor. It has a solubility of 1 g in approximately 10 ml of water with slight heating. It is used by the pharmaceutical industry as an iron supplement in vitamin pills. It is used by olive growers to darken the olives to a uniform black color. It can function as an iron fortifier in corn and soy products, breakfast cereals, beverages, and dietary foods.

**Ferrous Lactate**—Ferrous salt of lactic acid which functions to enrich and fortify. It is of neutral color and flavor, and is soluble. It is used in acid foods below a pH of 4.5, where there is less susceptibility to oxidation and conversion to the ferric form, which results in discoloration. In higher pH foods, the instability can be prevented by complexing with ligands which prevent browning discoloration. It is used in fortification of drinks and juices.

**Ferrous Sulfate**—A nutrient and dietary supplement that is a source of iron. It is a white to grayish odorless powder. Ferrous sulfate heptahydrate contains approximately 20 percent iron, while ferrous sulfate dried contains approximately 32 percent iron. It dissolves slowly in water and has high bioavailability. It can cause discoloration and rancidity. It is used for fortification of baking mixes. In the encapsulated form it does not react with lipids in cereal flours. It is used in infant foods, cereals, and pasta products.

**Fish Protein Isolate**—A food supplement that consists principally of dried fish protein prepared from the edible portion of fish after removal of the heads, fins, tails, bones, scales, viscera, and intestinal contents. The additive is prepared by extraction with hexane and food-grade ethanol to remove fat and moisture.

**Flavor Enhancers**—Substances added to supplement, enhance, or modify the original taste and/or aroma of a food, without imparting a characteristic taste or aroma of its own. *See Flavoring Agents and Adjuvants.*

**Flavoring Agents and Adjuvants**—Substances added to impart or help impart a taste or aroma in food. They are classified into the major groups of spices, natural flavors, and artificial or synthetic flavors. Aliphatic, aromatic, and terpene compounds refer to synthetic chemicals and isolates from natural sources. This classification encompasses the largest group of flavoring materials. The flavors used are natural, artificial, or combinations and exist in liquid or dry form. General flavor types available include fruit, dairy, meat, vegetable, beverage, and liquor.

**Flaxseed**—Seed from flax which is rich in omega-3 fatty acid (up to 24 percent by weight), total dietary fiber, and lignans. Flaxseed is also termed linseed. It is used in cereal-based products and cereal mix.

**Flour**—The food prepared by grinding and bolting cleaned wheat, other than durum wheat and red durum wheat. The baking quality of the flour depends upon the type of wheat, milling process, and treatment applied after milling. Flours classified by process are straight, patent, and clear flour. Flours classified by usage are all-purpose, bread, cake, cracker, and pastry flour. Flours treated after milling include bleached, bromated, enriched bleached, instantized, phosphated, and self-rising flour. Flours from other grains are identified according to the grain source, for example, soy flour. See specific flour.

**Flour Treating Agents**—Substances added to milled flour, at the mill, to improve its color and/or baking qualities, including bleaching and maturing agents.

**Foaming Agents**—*See Surface-Active Agents*.

**Folacin**—*See Folic Acid*.

**Folic Acid**—A water-soluble B-complex vitamin that aids in the formation of red blood cells, prevents certain anemias, and is essential in normal metabolism. High-temperature processing affects its stability. It is best stored at lower than room temperatures. It is also termed folacin. It is found in liver, nuts, and green vegetables.

**Food Starch, Modified**—*See Modified Starch*.

**Formic Acid**—A flavoring substance that is liquid and colorless, and possesses a pungent odor. It is miscible in water, alcohol, ether, and glycerin, and is obtained by chemical synthesis or oxidation of methanol or formaldehyde.

**Fructooligosaccharide**—(FOS) A fructan being a natural constituent of inulin and found in artichokes, asparagus, onions, garlic, and leeks. It is obtained from the partial enzymatic hydrolysis of inulin and can also be enzymatically synthesized from sucrose. Its properties include functioning as a prebiotic to promote growth of bifidogenic and other probiotic bacteria in the gastrointestinal tract, as a soluble dietary fiber, and to increase calcium absorption. In fruit preparations with aspartame or acesulfame potassium, it provides a synergistic taste affect and improves mouthfeel, it reduces stickiness in soy protein bars, binds water, and increases shelf life in baked goods.

**Fructose**—A sweetener that is a monosaccharide found naturally in fresh fruit and honey. It is obtained by the inversion of sucrose by means of the enzyme invertase and by the isomerization of corn syrup. It is 130 to 180 in sweetness range as compared to sucrose at 100 and is very water soluble. It is used in baked goods because it reacts with amino acids to produce a browning reaction. It is used as a nutritive sweetener in low-calorie beverages. It is also termed levulose and fruit sugar.

**Fructose Corn Syrup**—A sweetener that is an isomerized corn syrup derived from isomerization of glucose in the syrup to fructose by the enzyme isomerase. Varying levels of fructose syrup are available, being designated 42, 55, and 90 percent fructose. The 42 percent high-fructose corn syrup (HFCS) is a liquid mixture of dextrose, fructose, maltose, isomaltose, and higher saccharides, of which 42 percent is fructose (dry basis). The 55 percent and 90 percent HFCS are liquid mixtures of fructose, dextrose, and higher saccharides containing 55 percent and 90 percent fructose (dry basis), respectively. The range of relative sweetness as compared to sucrose at 100 is 42 percent HFCS: 90 to 95; 55 percent HFCS: 95 to 100; 90 percent HFCS: 100 to 130. HFCS is used in carbonated beverages, canned fruit, frozen desserts, and dairy drinks. It is also termed isomerized syrup, levulose-bearing syrup, and high-fructose corn syrup.

**Fruit Sugar**—*See Fructose*.

**Fumaric Acid**—An acidulant that is a nonhygroscopic, strong acid of poor solubility. It has a solubility of 0.63 g in 100 ml of distilled water at 25°C. It dissolves slowly in cold water, but if mixed with dioctyl sodium sulfosuccinate its solubility improves. The solubility rate also increases with smaller particle size. A quantity of 0.317 kg of fumaric acid can replace 0.453 kg of citric acid. It is used in dry mixes such as desserts, pie fillings, and candy. It is used in dry beverage mixes because it is storage stable, free flowing, and nonhygroscopic. It functions as a synergistic antioxidant with BHA and BHT in oil- and lard-base products. In gelatin desserts, it improves the flavor stability and increases shelf life and gel strength.

**Furcelleran**—A gum that is the extract of the red alga *Furcellaria fastigiata*. It swells in cold water and requires heating to 75 to 80°C for solubilization. It forms thermoreversible gels after heating and cooling and has properties between agar and carrageenan. It is also termed Danish agar. It is used in milk puddings, flans, jelly, jam, and gelled meat products.

# G

**Garlic**—A spice that is cloves of the herb *Allium sativum*. In its dehydrated form, the flavor enzyme is released only when in combination with water. It exists in powder form and also as salt, chips, and seasoning powder. It is used to flavor meats, vegetables, and sauces.

**Garlic Salt**—A seasoning that is a mix of garlic powder and salt. It is used in sauces and breads.

**Gelatin**—A protein that functions as a gelling agent. It is obtained from collagen derived from beef bones and calf skin (Type B) or pork skin (Type A). Type B is derived from alkali-treated tissue and has an isoelectric point between pH 4.7 and 5.0. Type A is derived from acid-treated tissue and has an isoelectric point between pH 7.0 and 9.0. It forms thermally reversible gels which set at 20°C and melt at 30°C. The gel strength is measured by means of a Bloom Gellometer and ranges from 50 to 300 with a 250 Bloom being the most common. It is used in desserts at 8 to 10 percent of the dry weight, in yogurt at 0.3 to 0.5 percent, in ham coatings at 2 to 3 percent, and in confectionery and capsules at 1.5 to 2.5 percent.

**Gellan Gum**— A gum obtained by fermentation of the microorganism *Sphingomonas elodea*. The constituent sugars are glucose, glucuronic acid, and rhamnose in the molecular ratio of 2:1:1, being linked together to give a primary structure consisting of a linear tetrasaccharine repeating unit. Direct recovery yields the gum in its native or high acyl form in which two acyl substituents, acetate and glycerate, are present. Gels from that form are elastic and cohesive. Recovery after deacetylation has the acyl groups removed to yield the low acyl form; those gels are strong and brittle. In general, high acyl gellan gum dispersed in water swells to form a thick suspension and upon heating, it loses its viscosity upon hydration. Low acyl gellan gum is only partially soluble in cold water and is dissolved by heating to 70°C or greater. Gelation occurs upon cooling and reaction with ions, predominantly calcium ions. Gellan gum is sensitive to ions. Uses include bakery fruit fillings, confectioneries, icings, dairy products, beverages, and coatings.

**Geranyl Isovalerate**—A synthetic flavoring agent that is a moderately stable, light yellow liquid of fruity odor. It should be stored in glass or tin containers. It is used in fruit flavors such as apple or pear with applications in beverages, ice cream, candy, and baked goods at 4 to 11 parts per million.

**Geranyl Phenylacetate**—A flavoring agent that is a yellow liquid with an odor resembling honey and roses. Miscible in alcohol, chloroform, and ether, and insoluble in water, it may contain other isomeric and closely related terpenic esters. It is obtained by chemical synthesis.

**Ghatti**—A gum that is a plant exudate obtained from the *Anogeissus latifolia* tree. The gum is formed as a protective sealant when the bark is damaged. It forms viscous mixtures in water at concentrations of 5 percent or greater. Only about 90 percent of the gum is actually soluble in water and has a pH of 4.5. It has similar uses as gum arabic. It is also termed Indian gum. It is used in buttered syrup and as a stabilizer for emulsions.

**Ghee**—*See Butter Oil*.

**Ginger**—A spice that is the dried and peeled rhizome of the ginger plant *Zingiber officinale*. The fragrance ranges from pungent to piquant at once; the flavor can be sharp or cooling depending on the food with which it is used. Fresh (green) ginger is obtained from the cleaned, peeled, and cured rhizome; dried ginger is the fresh product which has been cured and ground for spice. It is used in desserts, meats, sauces, relishes, baked goods, and beverages.

**Glacial Acetic Acid**—An acidulant that is a clear, colorless liquid which has an acid taste when diluted with water. It is 99.5 percent or higher in purity and crystallizes at 17°C. It is used in salad dressings in a diluted form to provide the required acetic acid. It is used as a preservative, acidulant, and flavoring agent. It is also termed acetic acid, glacial.

**Gluconic Acid**—An acidulant that is a mild organic acid which is the hydrolyzed form of glucono-delta-lactone. It is prepared by the fermentation of dextrose, whereby the physiological D-form is produced. It is soluble in water with a solubility of 100 g per 100 ml at

20°C. It has a mild taste and at 1 percent has a pH of 2.8. It functions as an antioxidant and enhances the function of other antioxidants. In beverages, syrups, and wine, it can eliminate calcium turbidities. It is used as a leavening component in cake mixes, and as an acid component in dry-mix desserts and dry beverage mixes.

**Glucono-Delta-Lactone**—(GDL) An acidulant. It hydrolyzes to form gluconic acid in water solution and thereby creates the desired pH. The rate of acid formation is affected by temperature, concentration, and the pH of the solution. It has low acid release at room temperature and accelerated conversion into gluconic acid at high temperatures. It is readily soluble with a solubility of 59 g in 100 ml of water at 20°C. It functions as a leavening agent, acidulant, curing and pickling agent, and pH control agent. It is comparatively less tart/sour than other food acids. It is used in baked goods, fish products, desserts, and dressings.

**Glucose**—*See Dextrose*.

**Glucose Syrup**—*See Corn Syrup*.

**Glutamic Acid**—An amino acid that is a white crystalline powder of slight solubility in water. The salt is monosodium glutamate (MSG) which functions as a flavor enhancer in meats. It also is a nutrient, dietary supplement, and salt substitute.

**Glutamic Acid Hydrochloride**—A flavoring, salt substitute that is soluble in water and very slightly soluble in alcohol and ether. It is obtained by chemical synthesis.

**Gluten**—A protein complex formed when water is kneaded with wheat flour which brings about the removal of a large portion of the starch. It forms the elastic framework of dough, entrapping the gas produced by the fermentation of leavening action which results in a risen dough of desired loaf volume and structure. Gliadin is of lower molecular weight and provides extensibility as compared to glutenin, which is of higher molecular weight and contributes elasticity. Gluten is available as wheat gluten, corn gluten, and zein. Vital wheat gluten is the most widely used. *See Wheat Gluten*.

**Gluten Flour**—*See Gluten*.

**Glycerin (Glycerol)**—A polyol (polyhydric alcohol) that functions as a humectant, crystallization modifier, and plasticizer. It is a bittersweet liquid which has a high solubility of 71 g per 100 g of water at 25°C. It is 75 percent as sweet as sugar. It is a fair oil solvent and has a medium to high hygroscopicity. It is used to maintain a certain moisture content to prevent the drying-out of foods; at 10 to 15 percent in raisins, it keeps them from drying out and prevents their moisture from migrating into cereal. It is used in confections to maintain the initial level of crystallization of the soft sugar. In reduced-fat frozen desserts, it helps prevent ice crystal formation. It also functions as a flavor solvent. Applications include marshmallows, candy, and baked goods.

**Glycerol—*See Glycerin*.**

**Glycerol Ester**—A density adjuster prepared from glycerol of non-animal sources and refined wood rosin of pine trees. It is used to adjust the specific gravity of the citrus oil or oil phase to be similar to the specific gravity of the beverage emulsion and thus prevent the oil from rising or settling in the finished beverage. It also imparts some cloudiness. It is soluble in aromatic and petroleum hydrocarbons, terpenes, esters, ketones, citrus, and essential oils. It is used in lemon and orange drinks and also as a masticatory substance in chewing gum base. It is technically termed glyceryl abietate and is also called glycerol dihydroabietate.

**Glyceryl-Lacto Esters of Fatty Acids**—Lipophilic emulsifiers that are the lactic acid esters of mono- and diglycerides. They are made by the reaction of mono- and diglycerides or propylene glycol ester with lactic acid, resulting in a compound with more surface activity and slightly more hydrophilicity than the regular mono- and diglycerides. They are used as emulsifiers, plasticizers, and promoters of starch gelatinization. They are used where aeration is required, such as in toppings, cakes, and icings, at levels necessary to obtain the technical effect.

**Glyceryl-Lacto-Stearate**—An emulsifier that is a glyceryl-lacto ester of fatty acids. It is a monoglyceride esterified with lactic acid which increases the hydrophilicity of the emulsifier. It is used in whipped vegetable toppings, shortenings, cake mixes, and chocolate coating.

**Glyceryl Monolaurate**—A monoglyceride emulsifier produced by the esterification of glycerine and lauric acid. It has a melting point of 56°C, a maximum iodine value of 0.5, and a saponification value of 200 to 206. In a highly purified form, it shows antimicrobial properties against microorganisms with the exception of gram-negative organisms. It is effective against gram-negative organisms when formulated with BHA or EDTA. It is used in baked goods, whipped toppings, frosting, glazes, and cheese products.

**Glyceryl Monooleate**—A flavoring agent that is prepared by esterification of commercial oleic acid that is derived either from edible sources or from tall oil fatty acids. It contains glyceryl monooleate and glyceryl esters of fatty acids present in commercial oleic acid. The ingredient is also used as an adjuvant and as a solvent and vehicle.

**Glyceryl Monostearate**—Glyceryl monostearate, also known as monostearin, is a mixture of variable proportions of glyceryl monostearate, glyceryl monopalmitate, and glyceryl esters of fatty acids present in commercial stearic acid. Glyceryl monostearate is prepared by glycerolysis of certain fats or oils that are derived from edible sources or by esterification, with glycerin, of stearic acid that is derived from edible sources.

**Glyceryl Triacetate**—A colorless, oily liquid of slight fatty odor and bitter taste. It is soluble with water and is miscible with alcohol and ether. It functions in foods as a humectant and solvent. It is also termed triacetin.

**Glyceryl Tristearate**—A formulation aid, lubricant, and release agent, prepared by reacting stearic acid with glycerol in the presence of a suitable catalyst. The additive is used as a crystallization accelerator in cocoa products; a formulation aid in confections; a formulation in fats and oils; and a winterization and fractionation aid in fat and oil processing.

**Glycine**—A nonessential amino acid that functions as a nutrient and dietary supplement. It has a solubility of 1 g in 4 ml of water and is abundant in collagen. It is used to mask the bitter aftertaste of saccharin, for example, in artificially sweetened soft drinks. It retards rancidity in fat.

**Glycyrrhizin**—A flavorant and foaming agent derived from the separation of flavonoids found in the whole licorice extract from the licorice root *Glycyrrhiza glabra*. It is 50 to 100 times as sweet as sugar, is soluble in water, and has a licorice taste. It has good heat stability but prolonged heating can result in some degradation. It is stable within pH 4 to 9; below pH 4 there could be precipitation. It has foaming and emulsifying properties in water, being used in cocktail mixes and soft drinks. It is used as a flavorant in bacon and imitation whipped products. It is synergistic with sugar, the sweetness being amplified to 100 times that of cane sugar alone. It is used as a sweetener in sugar-free chewing gum and low-fat sugar-free frozen desserts. It is also termed ammoniated glycyrrhizin. Monoammonium glycyrrhizinate is obtained by additional refinement.

**Golden Apple Seed—***See Quince Seed*.

**Graham Flour—***See Whole Wheat Flour*.

**Graham's Salt—***See Sodium Hexametaphosphate*.

**Grain Vinegar—**An acidulant made by the acetous fermentation of dilute distilled alcohol, containing not less than 4 g of acetic acid per 100 ml at 20°C. It is used in mayonnaise, salad dressing, sauces, and catsup. It is also termed distilled vinegar and spirit vinegar.

**Granulated Sugar—***See Sugar*.

**Grape Color Extract—**An aqueous solution of anthocyanin grape pigment made from Concord grapes, or a dehydrated water-soluble powder prepared from the aqueous solution. It contains the common components of the grape juice, but not in the same proportions. It has a red color pigment, with greatest color stability below pH 4.5. The color is stable in the presence of light and some heat. The color intensity increases as the pH declines. It is used at the 0.05 to 0.8 percent range. It may be used for coloring nonbeverage foods.

**Grape Seed Oil—**The oil obtained from grape seeds which contain an average of 15 percent oil. It is used as a drying oil with seeded raisins to improve their appearance and to prevent sticking. It is also termed raisin seed oil.

**Grape Skin Extract—**A natural red colorant with a high concentration of red anthocyanic pigments which provide its physicochemical properties. These pigments are responsible for the red, purple, violet, and blue hues of flowers and fruits. It is prepared by aqueous extraction of the fresh seedless marc remaining after the grapes have been pressed in the production of grape juice and wine. It contains the common components of grape juice, but in different proportions. The color depends upon the medium and the pH. In an acid medium and up to pH 4.5 to 5.5, the color is violet and becomes blue at pH 6.5. It has excellent water solubility and fair heat, light, and chemical stability. It can be used in soft drinks at 0.2 to 0.4 percent, and in candies at 0.5 to 1.5 percent.

**Guaiacol**—A precursor of vanillin and santalidol (a synthetic sandalwood fragrance). It is obtained from wood tar by the destructive distillation of hardwood, by the distillation of the phenol fraction of coal tar, or through the use of o-dichlorobenzene. It is processed to yield vanillin.

**Guar**—A gum that is a galactomannan obtained from the seed kernel of the guar plant *Cyamopsis tetragonoloba*. It is dispersible in cold water to form viscous sols which upon heating will develop additional viscosity. A 1 percent solution has a viscosity range of 2000 to 3500 centipoises. It is a versatile thickener and stabilizer used in ice cream, baked goods, sauces, and beverages at use levels ranging from 0.1 to 1.0 percent. It is scientifically termed guaran.

**Guaran**—*See Guar*.

**Gum Arabic**—*See Ghatti*.

**Gum Base**—The component of chewing gum that is insoluble in water and remains after chewing. It is prepared by blending and heating several ingredients to include a masticatory substance of vegetable or synthetic origin such as chicle, crown gum, petroleum wax, lanolin, polyethlyene, polyvinyl acetate, or rubber, with a plasticizer such as paraffin and with antioxidants. The gum base is 15 to 30 percent of chewing gum, of which a sweetener is the principal ingredient.

**Gum Ghatti**—*See Ghatti*.

**Gum Quince Seed**—*See Quince Seed*.

**Gums**—Polysaccharides that function as water-control agents by increasing viscosity (resistance to flow) or by forming gels. Gums are classified by source according to the following principal groupings: plant exudates, which include arabic, tragacanth, karaya, and ghatti; seaweed extracts, which include agar, alginates, carrageenan, and furcelleran; plant seed gums, which include guar, locust bean, tamarind, psyllium, and quince; plant extracts, which include pectin and arabinogalactan; fermentation gums, which include xanthan gum, gellan gum, and dextran; and cellulose derivatives, which include carboxymethyl cellulose, hydroxypropylmethyl cellulose, and microcrystalline cellulose. Gum derivatives include propylene glycol alginate and low-methoxyl pectin. They are also termed hydrocolloids.

**Gum Tragacanth**—*See Tragacanth*.

# H

**Heptanone**—A flavoring agent that is miscible in alcohol and ether, slightly soluble in water. It is obtained by chemical synthesis. This flavoring substance or its adjuvant may be safely used in food in the minimum quantity required to produce its intended flavor. It can be used alone or in combination with other legally approved flavoring substances or adjuvants. It is also termed methyl amyl ketone.

**Heptyl Cinnamate**—A synthetic flavoring agent that is a fairly stable, yellow liquid with a hyacinth odor. It should be stored in glass or tin containers. It is used to smooth out fruity flavors and has application in gelatins and puddings at approximately 20 parts per million and in candy, beverages, and ice cream at 2 to 6 parts per million.

**Heptyl Formate**—A synthetic flavoring agent that is a moderately stable, colorless to light yellow liquid of fruity odor. It should be stored in glass or tin containers. It is used in fruit flavors such as apricot, pear, and plum with applications in beverages, ice cream, candy, and baked goods at 1 to 4 parts per million.

**Heptyl Isobutyrate**—A synthetic flavoring agent that is a stable, colorless liquid of fruity odor. It should be stored in glass or tin containers. It is used in flavors for pineapple, pear, and orange with applications in beverages, ice cream, candy, and baked goods at 1 to 3 parts per million.

**Heptyl Paraben**—A preservative and antimicrobial agent. It is very slightly soluble in water. It may be used in fermented malt beverages to inhibit microbial spoilage and is permitted in beer. It is also termed N-heptyl-para-hydroxybenzoate.

**Hesperidin**—A flavoring agent that is a bioflavonoid found in citrus pulp. It has minor use as a flavorant.

**High-Fructose Corn Syrup**—(HFCS) A sweetener that is an isomerized corn syrup derived from the isomerization of the glucose in the syrup to fructose by the enzyme isomerase. Varying concentrations of fructose syrup are available, designated 42, 55, and 90 percent fructose. The 42 percent high-fructose corn syrup (HFCS) is a liquid mixture of dextrose, fructose, maltose, isomaltose, and higher saccharides, of which 42 percent is fructose, dry basis. The 55 and 90 percent HFCSs are liquid mixtures of fructose, dextrose, and higher saccharides containing 55 and 90 percent fructose, dry basis, respectively. The range of relative sweetness as compared to sucrose at 100 is 42 percent HFCS: 90 to 95; 55 percent HFCS: 95 to 100; 90 percent HFCS: 100 to 130. It is used in carbonated beverages, canned fruit, frozen desserts, and dairy drinks. It is also termed isomerized syrup, levulose-bearing syrup, and fructose corn syrup.

**Homogenized Milk**—Milk that has been mechanically treated to reduce the size of the fat globules such that after 48 hours of quiescent storage at about 7°C no visible cream separation occurs and the percentage of fat of the upper 100 ml in 946 ml of milk does not differ by more than 10 percent from the fat percentage of the remaining milk. Homogenization makes the milk more homogeneous but also decreases the heat stability of the milk proteins. It is used as a beverage and constituent of other food products. Practically all whole milk sold retail in the United States is homogenized.

**Honey**—A sweetener that is a natural syrup. It is similar to invert sugar, with a small but variable excess of levulose (fructose). It is formed by the action of the enzyme honey invertase on nectar gathered by bees. The composition and flavor varies with the plant source of the nectar, processing, and storage. A typical composition is 41 percent fructose, 34 percent glucose, 18 percent water, and 2 percent sucrose with a pH of 3.8 to 4.2. It is 1 to 1.5 times sweeter than sugar. It also functions to provide moisture, browning, and shelf life extension in some products. It is used in baked goods, cereals, and beverages.

**Horseradish**—A spice, the granules obtained from the horseradish plant. The flavor is released with moisture. It has a hot flavor character and has good stability. It is used in sauces.

**Hydrated Lime**—*See Calcium Hydroxide.*

**Hydrocolloids**—*See Gums.*

**Hydrochloric Acid**—An acid that is the aqueous solution of hydrogen chloride of varying concentrations. It is miscible with water and with alcohol. It is used as an acidulant and neutralizing agent.

**Hydrogenated Starch Hydrolysate**—Polyhydric alcohols (polyols) that do not contain a specific polyol as the majority component. Maltitol syrup is an example.

**Hydrogenated Vegetable Oil**—Oil that has been hydrogenated to modify the texture from a liquid to a semisolid or solid. The hydrogenization, which is the chemical addition of hydrogen, raises the melting point and converts the oil to a more desirable texture and consistency. It is used in farinaceous foods, confectionery, and desserts.

**Hydrolyzed Cereal Solids**—These are maltodextrins of low DE (dextrose equivalent). They function as anticaking agents, bodying agents, carriers, and crystallization inhibitors and are used in dry mixes and desserts.

**Hydrolyzed Protein**—*See Hydrolyzed Vegetable Protein.*

**Hydrolyzed Vegetable Protein**—(HVP) A flavor enhancer obtained from vegetable proteins such as wheat gluten, corn gluten, defatted soy flour, and defatted cottonseed. The proteins are hydrolyzed into their component amino acids after which the reaction mixture is neutralized with sodium carbonate and refined. The refined liquid HVP consists of amino acids, monosodium glutamate, amino acid derivatives, salt, and water. After being stored for several months, the liquid HVP is concentrated into a paste, dried, and ground. A typical dried HVP consists of 40 to 45 percent salt, which is generated during the neutralization process and serves to enhance the mouth feel of the HVP and provide preservation properties. It normally contains 9 to 12 percent monosodium glutamate and the remaining fraction consists of flavor solids. There are two basic types: pale HVP, which functions as a flavor enhancer with delicate spray flavors used in cream-type soups and sauces, and poultry; and dark HVP, which functions as a flavor donor with strong meaty flavors used in stews and broths. HVP is stable under varying processing conditions. It is used to improve flavors in soups, dressings, meats, snack foods, and crackers. It is also termed hydrolyzed protein.

**Hydroxylated Lecithin**—An emulsifier and clouding agent that is a modified crude lecithin of improved water dispersibility. It is manufactured by treating soybean lecithin with peroxide to increase the hydrophilic properties of lecithin. It is partially soluble in water but hydrates readily to form emulsions. It is used in bakery products because it has an apparent synergy with mono- and diglycerides. It is also used in dry-mixed beverages and margarine. It is also termed hydroxylated soybean lecithin.

**Hydroxylated Soybean Lecithin**—*See Hydroxylated Lecithin.*

**4-Hydroxymethyl-2,6-Di-Tert-Butylphenol**—An antioxidant used alone or in combination with other permitted antioxidants. The total amount of all antioxidants added to food must not exceed 0.02 percent of the oil or fat content of the food, including the essential (volatile) oil content of the food.

**Hydroxypropyl Cellulose**—A gum that is nonionic water-soluble cellulose, obtained from the reaction of alkali cellulose with propylene oxide at high temperatures and pressures. It is soluble in water below 40°C, is precipitated as a floc between 40 and 45°C, and is insoluble above 45°C. The precipitation is reversible with the original viscosity being restored upon cooling below 40°C and stirring. It is used in whipped toppings as a stabilizing and foaming aid; in edible food coatings as a glaze and oil/oxygen barrier; and in fabricated foods as a binder. Typical usage level is 0.05 to 1.0 percent.

**Hydroxypropyl Methylcellulose**—A gum formed by the reaction of propylene oxide and methyl chloride with alkali cellulose. It will gel as the temperature is increased in heating and upon cooling will liquefy. The gel temperature ranges from 60 to 90°C, forming semi-firm to mushy gels. It is used in bakery goods, dressings, breaded foods, and salad dressing mix for syneresis control, texture, and to provide hot viscosity. Usage level ranges from 0.05 to 1.0 percent.

# I

**Indian Gum**—*See Ghatti*.

**Indigotine**—*See FD&C Blue #2*.

**Indole**—A flavoring agent that is a white, flaky crystalline product. It has an unpleasant odor when concentrated and a flowery odor when diluted. It is soluble in most fixed oils and propylene glycol and insoluble in glycerin and mineral oil. It is obtained from decomposition of a protein.

**Instantized Flour**—A flour made by a milling or agglomerating procedure which makes it readily pourable, providing convenience.

**Inulin**—A non-digestible oligosaccharide containing fructose which provides texture, rheology, dietary fiber properties, and selective fermentation by colon bacteria. Commercially obtained from chicory root; common sources include onion, garlic, leek, asparagus, and Jerusalem artichoke. It is a hygroscopic powder with solubility in water dependent on water temperature. With increasing concentration, viscosity gradually increases, and at about 30 percent concentration, it can form discrete particle gels which are characterized as creamy and fat-like. It is not hydrolyzed by the digestive system. It functions as a prebiotic, passing into the colon where it is preferentially fermented by healthy bacteria such as bifidobacteria and lactobacilli to increase their proliferation and inhibit unwanted bacteria. It is used in ice cream products to replace fat and sugar, and in baked goods.

**Invert Sugar**—A sweetener that is a mixture of equal weights of dextrose (glucose) and levulose (fructose). It is more soluble than sucrose and has higher moisture-retaining properties because of the fructose content. It resists crystallization. It is used in candy and icings because it is sweeter, more soluble, and crystallizes less readily than sucrose.

**Invert Sugar Syrup**—A sweetener produced by an inversion process. It is produced by solubilizing sucrose in water followed by hydrolization to a mixture of dextrose and fructose using acids, invertase enzyme, or ion exchange resins to catalyze the reaction. Several invert syrups are obtained, such as medium invert consisting of 50 percent sucrose, 25 percent dextrose, 25 percent fructose; and total invert consisting of 3 to 5 percent sucrose, 48 percent dextrose, and 47 percent fructose. It has improved microbiological stability because of its high solids content, and it is used in soft drinks. It is also termed sugar syrup, invert.

**Iodine**—A halogen element extracted from Chilean nitrate-bearing earth or from seaweed. It functions by its presence in the thyroid hormones. Iodine deficiency is associated with goiter. Sources are potassium and cuprous iodide and potassium and calcium iodate, of which the iodate form is preferred because of better stability. It is used as a food supplement.

**Irish Moss**—A name sometimes used to denote carrageenan. It is a species of red seaweed known as *Chondrus crispus*, from which kappa and lambda carrageenans are obtained. *See Carrageenan*.

**Iron**—A mineral used in food fortification that is necessary for the prevention of anemia, which reduces the hemoglobin concentration and thus the amount of oxygen delivered to the tissues. Sources include ferric ammonium sulfate, chloride, fructose, glycerophosphate, nitrate, phosphate, pyrophosphate and ferrous ammonium sulfate, citrate, sulfate, and sodium iron EDTA. The ferric form ($Fe^{+3}$) is iron in the highest valence state and the ferrous form ($Fe^{+2}$) is iron in a lower valence state. The iron source should not discolor or add taste and should be stable. Iron powders produce low discoloration and rancidity. It is used for fortification in flour, baked goods, pasta, and cereal products.

**Iron Ammonium Citrate**—An anticaking agent used in salt.

**Iron–Choline Citrate Complex**—This special dietary additive is made by reacting approximately equimolecular quantities of ferric hydroxide, choline, and citric acid, and is used as a source of iron.

**Iron, Elemental**—A nutrient supplement, metallic iron is obtained by any of the following processes: reduced iron, electrolytic iron, and carbonyl iron.

**Iron Oxide**—A trace mineral used as a pigment and colorant. It is used to color pet food.

**Iron, Reduced**—Iron in a lower valence state, such as the ferrous form ($Fe^{+2}$). It is used in dry-mix oatmeal.

**Isoamyl Acetoacetate**—A synthetic flavoring agent that is a stable, colorless liquid of light green leaf–fruity odor. It should be stored in glass or tin containers. It is used in currant and berry flavors for applications in beverages, candy, and ice cream at 5 to 15 parts per million.

**Isoamyl Butyrate**—A synthetic flavoring agent that is a stable, colorless liquid of strong fruity odor. It is usually prepared by esterification of isoamyl alcohols with butyric acid. It is soluble in most fixed oils and mineral oil and is insoluble in glycerin and propylene glycol. Storage should be in glass, tin, or resin-lined containers. It is used in fruit flavors such as pineapple, raspberry, and strawberry and has application in dessert gels, puddings, and baked goods at 50 to 60 parts per million.

**Isoamyl Formate**—A synthetic flavoring agent that is a moderately stable, colorless to light yellow liquid of pungent pear-plum odor, being soluble in most fixed oils, mineral oil, and propylene glycol. Storage should be in a glass or tin container. It is used in fruit flavors such as pear, plum, and peach for application in dessert gels, puddings, candy, and ice cream at 14 to 28 parts per million.

**Isoamyl Hexanoate**—A synthetic flavoring agent that is a stable, colorless liquid of fruity odor. It is soluble in alcohol, fixed oils, and mineral oil. Storage should be in glass, tin, or resin-lined containers. It is used in fruit flavors such as banana and pineapple for applications in desserts, candy, and ice cream at 4 to 22 parts per million.

**Isoascorbic Acid**—*See Ascorbic Acid*.

**Isobutyl Acetate**—A flavoring agent that is a clear colorless liquid with a fruity odor resembling banana when diluted. It is soluble in alcohol, propylene glycol, most fixed oils, and mineral oil, and slightly soluble in water. It is obtained by synthesis.

**Isobutyl Cinnamate**—A synthetic flavoring agent that is a stable, colorless to light yellow liquid of fruity odor. It is miscible with alcohol, chloroform, and ether but is practically insoluble in water. Storage should be in glass or tin-lined containers. It is used in fruit flavors such as cherry and prune with applications in beverages, ice cream, candy, and baked goods at 1 to 5 parts per million.

**Isobutyl Formate**—A synthetic flavoring agent that is a stable, colorless liquid of fruity odor. Storage should be in glass or tin containers. It is used in fruit flavors such as pear, raspberry, and other berry flavors with applications in beverages, ice cream, candy, and baked goods at 2 to 18 parts per million.

**Isobutyric Acid**—A flavoring agent that is a colorless liquid with a strong, penetrating odor, resembling butter. It is miscible in alcohol, propylene glycol, glycerin, mineral oil, and most fixed oils and soluble in water. It is obtained by chemical synthesis. It is also termed isopropylformic acid.

**Isolated Soy Protein**—*See Soybean Protein Isolate*.

**Isomerized Syrup**—*See High-Fructose Corn Syrup*.

**Isopropyl Citrate**—An antioxidant that reacts with metal ions that might catalyze oxidative reactions and thus will prevent rancidity. It is made by reacting citric acid (not soluble in fats and oils) with isopropyl alcohol (which readily dissolves in oil) and thus enables the citrate to dissolve in oil. It is used in vegetable oils.

# J

**Juniper Berries Oil**—A flavoring agent that is a liquid which may be colorless, yellow or greenish in appearance. Its odor is characteristic with an aromatic, bitter taste. Storage is accompanied by polymerization. It is soluble in most fixed oils and mineral oil, insoluble in glycerin and propylene glycol. It is obtained from dried ripe fruit of *Juniperus communis* L. var. *erecta* Pursh of the *Cupressaceae* family.

**Karaya**—A gum, the dried exudate from the *Sterculia urens* tree which is native to India. It does not dissolve in water but swells to form a colloidal sol with a rate of hydration depending on mesh size. A 3 to 4 percent sol will result in a heavy gel and for higher concentrations the gum must be cooked under steam pressure to solubilize. It has a pH of 4.5 to 4.7. It functions as a binder and adhesive. It is used in baked goods, denture adhesives, toppings, and frozen desserts. It is also termed sterculia gum.

**Kelp**—A brown seaweed that grows in ocean water. The principal commercial species include *Macrocystis pyrifera* and *Laminaria hyperboria*. It is a source of alginic acid, which is used to produce alginate gum which functions as a water control agent. It contains the trace minerals potassium, sodium, calcium, and iodine. It is used as a source of iodine, as a flavor enhancer, as a nutrient and dietary supplement, and as a source of alginates.

**Kola Nut**—The seed of *Cola nitida* or other *Cola* species. The nut contains approximately 1.5 percent caffeine and is used in beverages and as an adjunct with other flavors.

# L

**Lactalbumin**—A milk protein obtained from whey by acidifying to pH 5.2, the isoelectric point, followed by coagulation by heat. It is not coagulated by rennin as in casein and is nonfunctional in its properties. It is used for nutritional purposes as a source of protein. It is used in cereals and breads where its relative inertness minimizes complications caused by other milk proteins during baking. It is also termed milk albuminate.

**Lactase Enzyme Preparation**—An enzyme preparation from *Kluyveromyces lactis* used to convert lactose to glucose and galactose. It is derived from the nonpathogenic, nontoxicogenic yeast *Kluyveromyces lactis* (previously named *Saccharomyces lactis*), and contains the enzyme B-galactoside galactohydrase, which converts lactose to glucose and galactose. It is prepared from yeast that has been grown in a pure culture fermentation and by using materials that are generally recognized as safe or food additives that have been approved for this use. This ingredient is used in milk to produce lactase-treated milk, which contains less lactose than regular milk, or lactose-reduced milk, which contains at least 70 percent less lactose than regular milk.

**Lactate**—Salts of lactic acid (calcium, sodium, aluminum, ammonium, ferrous, potassium, magnesium, manganese, and zinc) used to enrich and fortify diet foods, drinks, and juices.

**Lactic Acid**—An acidulant that is a natural organic acid present in milk, meat, and beer, but is normally associated with milk. It is a syrupy liquid available as 50 and 88 percent aqueous solutions, and is miscible in water and alcohol. It is heat stable, nonvolatile, and has a smooth, milk acid taste. It functions as a flavor agent, preservative, and acidity adjuster in foods. It is used in Spanish olives to prevent spoilage and provide flavor, in dry egg powder to improve dispersion and whipping properties, in cheese spreads, and in salad dressing mixes.

**Lacticol**—A polyhydric alcohol (polyol) derived from lactose by catalytic hydrogenation. It is 30 to 40 percent as sweet as sucrose. It has a taste profile and solubility that is comparable to sugar. Uses include bakery products, chewing gum, and tablets.

**Lactoglobulin**—A protein that is a complex of closely related proteins known as beta-globulins obtained from the whey fraction of milk. It is crystallizable and heat-denaturable.

**Lactose**—A disaccharide carbohydrate that occurs in mammalian milk except that of the whale and the hippopotamus. It is principally obtained as a cows' milk derivative. It is also termed milk sugar and it is a reducing sugar consisting of glucose and galactose. Its most common commercial form is alpha-monohydrate, with the beta-anhydride form available to a lesser extent. All forms in solution will equilibrate to a beta : alpha ratio of 62.25 : 37.75 at 0°C. It is about one-sixth as sweet as sugar and is less soluble. It functions as a flow agent, humectant, crystallization control agent, and sweetener. It is used in baked goods for flavor, browning, and tenderizing and in dry mixes as an anticaking agent.

**Lactylated Fatty Acid Esters of Glycerol and Propylene Glycol**—An emulsifier made by the reaction of a propylene glycol ester with lactic acid. It has more surface activity and is slightly more hydrophilic than mono- and diglycerides. It is used mainly where aeration is required, such as in toppings, cake mixes, and icings. It is used at levels required to produce the intended effect, such as 0.6 percent in fluid whipped topping and 0.5 percent in coffee whitener.

**Lactylic Esters of Fatty Acids**—An emulsifier that is mixed fatty acid esters of lactic acid and its polymers. It is dispersible in water and soluble in organic solvents and vegetable oils. It functions as a foaming agent in starch/protein systems and is used in puddings and coffee whiteners.

**Larch Gum**—*See Arabinogalactan.*

**Lard**—A fat rendered from hogs, consisting principally of oleic and palmitic fatty acids. It has a Wiley melting point of 88 to 110°F. It is rapidly chilled, resulting in an opaque, firm consistency rather than a translucent, greasy appearance. It is used in cake mix.

**Lauric Acid**—A fatty acid obtained from coconut oil and other vegetable fats. It is practically insoluble in water but is soluble in alcohol, chloroform, and ether. It functions as a lubricant, binder, and defoaming agent.

**Leavening Agents**—Acidic agents that chemically react with alkaline sodium bicarbonate to produce carbon dioxide gas. This reaction is initiated by moisture and completed by heat as the prepared mixture is baked. The value of the leavening agent relates to the rate upon which carbon dioxide is released from sodium bicarbonate, the suitability of the release rate to the product, and the mixing-raising-baking cycle. Leavening agents include tartaric acid, monocalcium phosphate, sodium acid pyrophosphate, sodium aluminum phosphate, and acidic acid.

**Lecithin**—An emulsifier that is a mixture of phosphatides which are typically surface-active. It is now commercially obtained from soybeans; previously it was obtained from egg yolk. It is used in margarine as an emulsifier and antispatter agent; in chocolate manufacture it controls flow properties by reducing viscosity and reducing the cocoa butter content from 3 to 5 percent; it is used as a wetting agent in cocoa powder, fillings, and beverage powders; an antisticking agent in griddling fat; and in baked goods to assist the shortening mix with other dough ingredients and to stabilize air cells. Typical usage levels range from 0.1 to 1.0 percent.

**Lecithinated Soy Flour**—Soy flour to which lecithin is added. The lecithin contributes emulsification and pan release properties. It is used in breading, caked foods, and dough mixes.

**Lemon Oil**—A flavoring agent that is the oil obtained from lemon fruit. It is used to impart lemon flavor and is used in reconstituted lemon juice.

**Levulose**—*See Fructose*.

**Levulose-Bearing Syrup**—*See High-Fructose Corn Syrup*.

**Licorice**—A flavoring agent made from dried root portions of *Glycyrrhiza glabra*. The obtainable forms are licorice root, licorice extract powder, and licorice extract. The extract is used in candy, baked goods, and beverages; the major licorice use is in tobacco.

**Lime**—*See Calcium Oxide*.

**Limestone**—*See Calcium Carbonate*.

**Limonene**—An antioxidant and flavoring agent that occurs in lemons, oranges, and pineapple juice, being obtained from the oils. It is a colorless liquid which is insoluble in water and propylene glycol, very slightly soluble in glycerine, and miscible with alcohol, most fixed oils, and mineral oil. It prevents or delays enzymatic browning-type oxidation.

**Linalyl Isobutyrate**—A flavoring agent that is a liquid, slightly yellow in color with a fruity odor. It is miscible in alcohol, ether, and chloroform, and insoluble in water. It is obtained by chemical synthesis. It is also termed 3,7-dimethyl-2,6-octadien-3-yl isobutrate.

**Locust Bean Gum**—A gum that is a galactomannan obtained from the plant seed from the locust bean tree known as *Ceratonia siliqua*. Its properties include swelling partially in cold water but requiring heating to approximately 82°C for complete solubility. It provides high viscosity, forms gels with xanthan gum upon heating and cooling of the solution, and functions as a water binder. It can make agar or carrageenan gels more elastic. Its uses include processed cheese, ice cream, bakery products, soups, and pies. Typical usage level is 0.1 to 1.0 percent. It is also called carob gum or Saint John's bread, and is scientifically called carubin.

**Low-Methoxyl Pectin**—A gum derived from pectinic acid. It differs from pectin in having a lower degree of methylation, less than 50 percent. It is also not as sensitive to pH and does not require sugar for gel formation. It forms thermally reversible gels with calcium salts and boiling may be required for solubility if the methoxyl content is low. It is used in low-calorie jellies at levels of 0.8 to 1.4 percent, in dessert and bakery jellies at levels of 1 to 1.5 percent, and in fruit gels at levels of 0.5 to 1.4 percent.

**M**

**Mace**—A spice that is the aril or skin covering of the nutmeg *Myristica fragrans* Houtt. It is more pungent in flavor than nutmeg. The whole mace is used in cooked fruit, pickles, and preserves, while ground mace is used in breads, cakes, chocolate pudding, and fruit salad.

**Magnesium**—A metallic element that is involved in certain bodily functions. Sources of magnesium include magnesium chloride and magnesium oxide. It functions as a nutrient and dietary supplement.

**Magnesium Carbonate**—An anticaking agent and general purpose food additive. It is practically insoluble in water but is more soluble in water containing carbon dioxide. It imparts a slightly alkaline reaction to the water. It is used as an alkali in sour cream, butter, and canned peas. It is used as an anticaking agent in table salt and dry mixes. It assists in providing clarity in algin gels and functions as a filler in dental impression materials.

**Magnesium Caseinate**—The magnesium form of caseinate which is marginal in functionality as compared to other forms of caseinates. It can be used in bakery goods, drinks, and dietary applications. ***See Caseinates***.

**Magnesium Chloride**—A source of magnesium, a color-retention agent, and firming agent. It exists as colorless flakes or crystals and is very soluble in water.

**Magnesium Hydroxide**—An alkali that is a general purpose food additive. It exists as a white powder and has poor solubility in water and in alcohol. In frozen desserts it will increase the tendency for fat globules to clump, which results in an increase in dryness. It reacts with triglycerides in fatty acids to form soaps. It also functions as a drying agent in foods.

**Magnesium Laurate**—The magnesium salt of lauric acid which functions as a binder, emulsifier, and anticaking agent.

**Magnesium Myristate**—The magnesium salt of myristic acid which functions as a binder, emulsifier, and anticaking agent.

**Magnesium Oleate**—The magnesium salt of oleic acid which functions as a binder, emulsifier, and anticaking agent.

**Magnesium Oxide**—A source of magnesium which functions as a nutrient and dietary supplement. It exists as a bulky white powder termed light magnesium oxide or as a dense white powder known as heavy magnesium oxide. It is practically insoluble in water and is insoluble in alcohol.

**Magnesium Palmitate**—The magnesium salt of palmitic acid which functions as a binder, emulsifier, and anticaking agent.

**Magnesium Silicate**—A white powder that is insoluble in water and functions as an anticaking agent. It is used in salt. It also is a processing aid and adsorbent which functions as an anticaking agent and remover of undesirable proteins during filtration. It is insoluble and a 10 percent slurry has a pH of approximately 7.0. It aids in the processing of beverages, food products, and pharmaceuticals by removing protein/tannin complex constituents through surface area and adsorptive effects.

**Magnesium Stearate**—The magnesium salt of stearic acid which functions as a lubricant, binder, emulsifier, and anticaking agent. It is a white powder that is insoluble in water. It is used as a lubricant or die release in tableting pressed candies and is also used in sugarless gum and mints.

**Magnesium Sulfate**—A nutrient and dietary supplement. It is soluble in water and its solutions are neutral. It exists as crystals with a cooling, saline, bitter taste. It is also termed Epsom salt.

**Maize Meal**—The meal obtained by grinding maize (Indian corn).

**Maize Starch**—*See Cornstarch.*

**Malic Acid**—An acidulant that is the predominant acid in apples. It exists as white crystalline powder or granules and is considered hygroscopic. As compared to citric acid, it is slightly less soluble but is still readily soluble in water with a solubility of 132 g per 100 ml at 20°C. It has a stronger apparent acid taste and has a longer taste retention than citric acid which peaks faster but does not mask the aftertaste as effectively. A quantity of 0.362 to 0.408 kg of malic acid is equivalent to 0.453 kg of citric acid and to 0.272 to 0.317 kg of fumaric acid in tartness. At temperatures above 150°C it begins to lose water very slowly to yield fumaric acid. It is used in soft drinks, dry-mix beverages, puddings, jellies, and fruit filling. It is used in hard candies because it has a lower melting point (129°C) than citric acid which improves the ease of incorporation.

**Malt**—A source of the enzyme alpha-amylase which hydrolyzes starch to fermentable sugars such as dextrins and maltose. It is produced by the controlled sprouting of grains, usually barley, followed by drying to produce three general classes of malt differing in amylase content. These classes are brewer's malt, distiller's malt, and gibberellin malt. Malt is used in the brewing industry and as a supplement to flour to increase the alpha-amylase content.

**Malted Barley**—The barley produced under the controlled sprouting of the barley grain followed by drying to obtain the formation of high levels of alpha-amylase and some increase in beta-amylase. These enzymes hydrolyze starch to dextrins and maltose. There are three general classes of malt: brewer's malt, distiller's malt, and gibberellin malt. It is principally used in the brewing industry. In doughs, the malt increases the fermentation rate and improves baking properties.

**Malted Cereal Syrup**—The syrup obtained from barley and other grains, as differentiated from malt syrup which is obtained only from barley. It is used to contribute flavor in baked goods and is a nutrient in yeast fermentation. It is also termed extract of malted barley and corn.

**Malt Extract**—A flavorant formed by extracting the water-soluble enzymes from barley and evaporating to form a concentrate that contains *D*-alpha-amylase enzyme. This enzyme hydrolyzes starch to dextrin and maltose. It is used to provide malt flavor, and in conjunction with spices, seasonings, and flavors.

**Malt Flour**—The flour prepared by the drying and grinding of barley or wheat sprouted under controlled conditions. It can be used as a malt supplement or converted to malt syrups. The malt functions to modify starch during initial stages of baking.

**Maltitol**—A polyhydric alcohol (polyol) produced by hydrogenation of maltose. It is approximately 90 percent as sweet as sucrose, has good stability, and is nonhygroscopic. Uses include chewing gum, dry nut bakery products, and chocolate.

**Maltodextrin**—Short-chain saccharide polymers obtained from the partial acid or enzymatic hydrolysis of starch, in the same manner as corn syrup except the conversion process is stopped at an earlier stage. It consists of *D*-glucose units linked principally by alpha-1,4 bonds, has a dextrose equivalent of less than 20 and basically is not sweet and is not fermentable. It has fair solubility. It functions as a bodying agent, bulking agent, texturizer, carrier, and crystallization inhibitor. It is used in crackers, puddings, candies, and sugar-free ice cream.

**Maltol**—A flavor enhancer used as a synthetic flavoring substance, the function of which is related to ethyl maltol. It occurs naturally in chicory, cocoa, coffee, and cereals. It does not contribute a flavor of its own, but modifies the inherent flavors. As compared to ethyl maltol, it is one-half to one-sixth as effective. It is less soluble, having a solubility of 1 g in 82 ml of water at 25°C. It has a melting range of 160 to 164°C. It is used to enhance the flavor and aroma of fruit, vanilla, and chocolate flavored foods and beverages. It is also used in beverages and desserts with a typical usage range of 10 to 200 parts per million.

**Maltose**—A sweetener formed by the enzymatic action of yeast on starch. It consists of two dextrose molecules. Maltose dissolves and crystallizes slowly in aqueous solutions, and is less sweet and more stable than sucrose. It is used in combination with dextrose in bread and in instant foods, and is also used in pancake syrups.

**Malt Syrup**—The syrup obtained from barley by extraction and evaporation of the worts to 80 to 81 percent solids. It is used as a malt flavor component, as a source of malt and protein, and to provide color. It is used in bakery goods such as rolls and bagels at 1 to 3 percent of the flour weight, in soybean milk at 3 to 7 percent, and in malt base at 1 to 3 percent.

**Malt Vinegar**—A vinegar made by the alcoholic and subsequent acetous fermentation of an infusion of malted barley and/or cereals or a concentrate thereof, which has been enzymatically converted by the malting process. It contains a minimum of 4 percent acid content expressed as acetic acid and is brown to dark brown in color. It functions as an acidulant and preservative in foods.

**Manganese**—A metallic element that functions as a nutrient and dietary supplement. It is necessary for normal bone and tendon structure, central nervous system functionality, and glucose utilization. Sources include manganese carbonate, chloride, oxides, and sulfates. These sources differ in solubility.

**Manganese Chloride**—A source of manganese that functions as a nutrient and dietary supplement. It exists as crystals which are readily soluble in room temperature (22°C) water and are very soluble in hot water. *See Manganese.*

**Manganese Citrate**—A nutrient supplement that is a pale orange or pinkish white powder. It is obtained by precipitating manganese carbonate from manganese sulfate and sodium carbonate solutions. The filtered and washed precipitate is digested first with sufficient citric acid solutions to form manganous citrate and then with sodium citrate to complete the reaction. It is used in baked goods, nonalcoholic beverages, dairy product analogs, fish products, and poultry products. The ingredient may be used in infant formulas.

**Manganese Gluconate**—A nutrient supplement that is a slightly pink-colored powder. It is obtained by reacting manganese carbonate with gluconic acid in aqueous medium and then crystallizing the product. The supplement is used in baked goods, nonalcoholic beverages, dairy product analogs, fish products, meat products, milk products, and poultry products. The ingredient may be used in infant formulas.

**Manganese Sulfate**—A source of manganese that functions as a nutrient and dietary supplement. It exists as a powder which is readily soluble in water. *See Manganese.*

**Mannitol**—A polyol (polyhydric alcohol) produced from hydrogenation from frutose that functions as a sweetener, humectant, and bulking agent. It has low hygroscopicity and poor oil solvency. It has 1.6 Kcal/gram. It is approximately 22 percent soluble in water and is approximately 72 percent as sweet as sugar, exhibiting a cool, sweet taste. It functions as a dusting agent with starch in chewing gum. It is used in sugarless candy, chewing gum, cereal, and pressed mints.

**Maple Sugar**—A sweetener obtained by concentrating the sap of the maple sugar tree. It consists of approximately 95 percent sucrose, 2 percent invert sugar, and ash. This is the dry form of maple syrup which is used in syrups and candy.

**Maple Syrup**—A sweetener made by concentrating the sap of the sugar maple tree by boiling at atmospheric pressure. The characteristic color and flavor are developed by heating above 100°C. The concentration at reduced pressure or by freeze-drying gives a sweet, colorless syrup. The characteristic flavor is derived from the volatile oil in the sap. On a dry basis it is approximately 95 percent sucrose, 2 percent invert sugar, and ash. It is used in syrups and candies.

**Margarine**—A product whose consistency and application are similar to those of butter. It is made by emulsifying vegetable oils with milk, followed by crystallization and kneading. Vegetable oils or mixtures of vegetable oils and animal fat may be used. It contains not less than 80 percent fat and is also termed oleomargarine. It is used as a spread and as a source of fat for baked goods and desserts.

**Marjoram**—A spice that is the dried leaves of the herb *Majorana hortensis* Moench. It has a mellow flavor and is distinctively aromatic. The flavor is close to that of oregano. Marjoram is used in soups, sauces, meats, and fish.

**Menhaden Oil**—A fish oil which is a source of omega-3 polyunsaturated fatty acids associated with nutritional and health benefits. It is generally recognized as safe with specific limitations. Uses include eggs, baked goods, salad dressings, and health supplements.

**Methacrylic Acid–Divinylbenzene Copolymer**—A carrier of vitamin $B_{12}$ in foods for special dietary use, produced by the polymerization of methacrylic acid and divinylbenzene. The divinylbenzene functions as a cross-linking agent and constitutes a minimum of 4 percent of the polymer.

**Methyl B-Naphthyl Ketone**—A flavoring agent that is a crystalline solid (white or nearly white) with orange blossom-like odor. It is soluble in most fixed oils, slightly soluble in mineral oil and propylene glycol, and insoluble in glycerin. It is obtained by chemical synthesis. It is also termed 2'-acetonaphtone.

**Methylcellulose**—A gum composed of cellulose in which the methoxyl groups replace the hydroxyl groups. It is soluble in cold water but insoluble in hot water. Solutions increase in viscosity upon heating, gel at 50 to 55°C, and liquefy upon cooling. It is used in baked goods for moisture retention, and in fruit pie fillings for the reduction of water absorption into the pie crust during baking. It is also used in breaded shrimp where it functions to form an oil barrier film.

**Methyl Cyclopentenolone**—A flavoring agent that is a white crystalline powder. It has a nutty odor suggesting a maple-licorice aroma when diluted. It is soluble in alcohol and propylene glycol, slightly soluble in most fixed oils, and sparingly soluble in water. It is obtained by synthesis. It is also termed 3-methyl-cyclopentane-1,2-dione.

**Methyl Ethyl Cellulose**—An aerating, emulsifying, and foaming agent. The methoxy content should be not less than 3.5 percent and not more than 6.5 percent, and the ethoxy content should be not less than 14.5 percent and not more than 19 percent, both measured on the dry sample.

**Methyl 2-Methylthiopropionate**—A synthetic flavoring agent that is a colorless liquid of slightly fruity odor with a suggestion of sulfur. It polymerizes slowly and should be stored in glass or tin containers. It is used in pineapple flavors to give the true note of pineapple. It has applications in beverages, ice cream, candy, and baked goods at 0.5 to 1 part per million.

**Methylparaben**—An antimicrobial agent which is a white free-flowing powder. It is active against yeast and molds over a wide pH range. *See Parabens*.

**Methyl p-Hydroxybenzoate**—*See Parabens*.

**Methyl Polysilicone**—*See Dimethylpolysiloxane*.

**Methyl Silicone**—*See Dimethylpolysiloxane*.

**3-Methylthiopropionaldehyde**—A synthetic flavoring agent that is a colorless to light yellow liquid with an intense meat odor. It polymerizes with age and is stable in a 50 percent alcohol solution. It should be stored in glass containers. It is used at low concentrations for meat and broth flavors for applications in meats and condiments at 3 parts per million and in baked goods and beverages at 0.5 part per million.

**Microcrystalline Cellulose**—A gum that is the nonfibrous form of cellulose, an alpha-cellulose. It is dispersible in water but not soluble, requiring considerable energy to disperse and hydrate. In this form it is used in dry applications such as tableting, capsules, and shredded cheese where it functions as a non-nutritive filler, binder, flow aid, and anticaking agent. By addition of carboxymethylcellulose to the alpha-cellulose prior to drying, improved functional properties of hydration and dispersion are obtained. This product is designed for use in water dispersions, being insoluble in water but dispersing in water to form colloidal sols below 1 percent and white opaque gels above the 1 percent usage level. It is used as a heat shock stabilizer and bodying agent in frozen desserts, as an opacifier in low-fat dressings, as a foam stabilizer in whipped toppings, and as an emulsifier in dressings.

**Microparticulated Protein Product**—A fat replacer prepared from egg whites or milk protein or combination egg whites and milk protein. These protein sources may be used alone or in combination with other safe and suitable ingredients to form the microparticulated product. The mixture of ingredients is high-shear heat processed to achieve a smooth and creamy texture similar to that of fat. The ingredient is used in food as a thickener or as a texturizer. It is used in frozen desserts, cheese, dressings, baked goods, and dairy products.

**Milk**—The natural secretion of the mammary glands of female mammals for the feeding of their young. It is commercially considered here as cows' milk which consists, on the average, of 3.5 percent fat, 5 percent lactose, 3.5 percent protein, and 0.7 percent ash. It has a bland, slightly sweet flavor, a yellowish white color, and a specific gravity of 1.032. It functions as a base for ice cream, yogurt, beverages, and cheese. It is also the source of skim milk, cream, whey, casein, lactose, and milk solids–not-fat. It is also termed whole milk.

**Milk Albuminate—See Lactalbumin.**

**Milk Chocolate—See Chocolate.**

**Milkfat**—The fat of milk which exists in milk as an emulsion of small fat globules in an aqueous phase. It is the only fat in which butyric acid is a component of the glycerides. It has a delicate and pleasant flavor. Approximately 95 percent of the total milk lipids are triglycerides. The average fat content of milk is 3.5 to 3.8 percent. It is used as a source of fat in bakery products, confections, and frozen desserts. It is also termed butter fat.

**Milkfat, Anhydrous—See Butter Oil.**

**Milk Powder—See Whole Milk Solids.**

**Milk Solids–Not-Fat**—The dry form of skim milk. It contains not more than 1.5 percent fat and not more than 5 percent moisture. It has excellent flavor, nutritional value, and functional properties such as water binding, emulsification, and foam formation. It is used in ice cream mix, baked goods, and desserts. It is also termed nonfat dry milk, skim milk powder, and dried skim milk.

**Milk Sugar—*See Lactose*.**

**Mint**—A spice derived from any one of the plants of the mint family (*Labiatae*) of which there are numerous varieties. Only spearmint and peppermint are commercially important. Mint is used in mint sauce, fruit cocktails, and beverages in either its dried or fresh forms.

**Modified Food Starch—*See Modified Starch*.**

**Modified Hop Extract**—A flavoring agent in the brewing of beer. It is manufactured from a hexane extract of hops, with simultaneous isomerization and selective reduction in an alkaline aqueous medium with sodium borohydride. It is added to the wort before or during cooking in the manufacture of beer.

**Modified Starch**—The product resulting from the treatment of starch with certain chemicals to modify the physical characteristics of the native starch. This produces more desirable or useful characteristics such as improved solubility, acid stability, and texture. It is used in desserts, pie fillings, sauces, gravies, and fabricated foods as a thickener, binder, and stabilizer. It is also termed modified food starch.

**Molasses**—The byproduct of the manufacture of sugar from sugar cane in which the syrup is separated from the crystals. The highest grade is edible molasses which is most often found as table syrup. The lowest grade is blackstrap molasses. Molasses is a strongly flavored, dark colored syrup containing 70 to 80 percent solids of which 50 to 75 percent is sucrose and invert sugar. It is used in syrups and in the production of caramel.

**Monoammonium Glutamate—*See Monoammonium L-Glutamate*.**

**Monoammonium Glycyrrhizinate**—A flavoring and sweetening agent obtained from licorice flavonoids. It has a slight licorice flavor and an intense and rounded sweetness. It is soluble in water, glycerin, and propylene glycol and has good thermal stability but can degrade under prolonged heating. It is stable over a pH of 3.5 to 9 but could precipitate below pH of 3.5. It can be used in beverages, desserts, confectionary products, soups, and dressings. ***See Glycyrrhizin***.

**Monoammonium L-Glutamate**—A flavor enhancer and salt substitute that is crystalline powder (white, free-flowing) and odorless.

It is soluble in water, insoluble in common organic solvents, and is obtained by chemical synthesis. It is a low sodium alternative to monosodium glutamate. It is used in meats, soups, gravies, and sausage. It is also termed ammonium glutamate and monoammonium glutamate.

**Monocalcium Phosphate**—An acidulant, leavening agent, and nutritional supplement that exists as white crystals or granular powder. It is sparingly soluble in water. It is used as an acidulant in breads and dry-mix beverages; as a source of calcium in fruit jellies, preserves, and cereals; and as a component of dough conditioners. It is also of restricted use as a chemical leavening agent because it releases about 67 percent of the carbon dioxide during the initial mixing and this is generally too rapid. It is also termed calcium acid phosphate, calcium phosphate monobasic, calcium biphosphate, and acid calcium phosphate.

**Monoglyceride**—A lipophilic emulsifier prepared by the direct esterification of fatty acids with glycerol or by the interesterification between glycerol and other triglycerides. It is insoluble in water. It provides emulsion stability, prevents fat separation, and also functions as a foaming agent, defoaming agent, and dispersant. It is most often used in combination with a diglyceride. It is used in ice cream, peanut butter, puddings, and numerous other applications. The typical usage level is 0.05 to 0.40 percent.

**Mono- and Diglycerides**—A lipophilic emulsifier that consists of both monoglycerides and diglycerides. It is made by reacting glycerol with specific fats or oils. The consistency varies from yellow liquid to ivory-colored plastic to hard solids of bland odor and taste. It is the most commonly used emulsifier in foods. It is used in numerous applications, for example, in baked goods, frozen desserts, whipped toppings, and margarine for a variety of functions. Typical usage levels range from 0.05 to 0.50 percent. It is also termed monosodium phosphate derivatives of mono- and diglycerides.

**Monoglyceride Citrate**—A sequestrant that is a mixture of glyceryl monooleate and its citric acid mono-ester. It is soluble in common fat solvents and alcohol and is insoluble in water. It is used in antioxidant formulations for addition to fats and oils at less than 200 parts per million. It functions as an antioxidant synergist in peanut oil at a maximum usage level of 100 mg/kg individually or in combination. In margarine, it is used at no more than 0.01 percent.

**Monoglyceride, Distilled**—*See Distilled Monoglyceride*.

**Monoisopropyl Citrate**—A sequestrant used in fats and oils.

**Monopotassium Glutanate**—A flavor enhancer that is a low sodium alternative to monosodium glutamate. It is used in meats, soups, sauces, gravies, and sausage.

**Monopotassium Monophosphate**—*See Monopotassium Phosphate*.

**Monopotassium Phosphate**—A buffer, neutralizing agent, and sequestrant. It is mildly acid, with a pH of 4.5, and fairly soluble in water, with a solubility of 25 g in 100 ml at 25°C. It is used in whole eggs for color preservation and is also used in low-sodium products, milk products, and meat products. Typical usage ranges from 0.1 to 0.5 percent. It is also termed potassium dihydrogen orthophosphate, potassium phosphate monobasic, and monopotassium monophosphate.

**Monosodium Dihydrogen Orthophosphate**—*See Monosodium Phosphate*.

**Monosodium Glutamate**—(MSG) A flavor enhancer that is the sodium salt of glutamic acid, an amino acid. It is a white crystal that is readily soluble in water. It intensifies and enhances flavor but does not contribute a flavor of its own. It may be present as one of the amino acids or in a free form, which is how it effectively enhances the flavor of foods. It is produced through a fermentation process of molasses. It is used at 0.1 to 1.0 percent in meats, soups, and sauces.

**Monosodium Monophosphate**—*See Monosodium Phosphate*.

**Monosodium Phosphate**—An acidulant, buffer, and sequestrant that is mildly acid, with a pH of 4.5, and very soluble in water, with a solubility of 87 g per 100 ml of water at 25°C. It is used as an acidulant in effervescent powders and laxatives. It is also used in soft drink dry-mix formulations, in cheese, and in carbonated beverages. It is also termed monosodium dihydrogen orthophosphate; sodium phosphate, monobasic; sodium biphosphate; sodium acid phosphate; and monosodium monophosphate.

**Monosodium Phosphate Derivatives of Mono- and Diglycerides**—*See Mono- and Diglycerides*.

**Mono-Tertiary-Butylhydroquinone**—*See Tertiary Butylhydroquinone*.

**Mustard**—A flavorant made from the dried, ripe seed of several closely related genera, species, and varieties of the family *Cruciferae*; the seed of a plant of the cabbage family. It is used as a flavorant in baked goods, sauces, and salad dressings. It also functions as an emulsifier in salad dressings. The ground seed is used for spices.

**Mustard Flour**—The ground seed of the mustard plant from which some of the oil and most of the hulls have been removed. It is used in salad dressings and sauces, and as a condiment.

**Mustard Oil**—*See Allyl Isothiocyanate*.

**Mustard Seed**—A spice of which there are several varieties, the dry mustards being of the hot or mild type. It is used in meats, sauces, and salad dressings.

**Myristic Acid**—A fatty acid obtained from coconut oil and other fats. It has poor water solubility but is soluble in alcohol, chloroform, and ether. It is used as a lubricant, binder, and defoaming agent.

# N

**Natamycin**—A preservative for use as a coating on the surface of Italian cheeses to prevent the growth of mold or yeast. It is tasteless, odorless, colorless, and does not penetrate the cheese. It is very active against virtually all molds and yeasts, but does not affect bacteria, thus not affecting the ripening and flavor improvement process of cheese. It can be applied as a dip, spray, or by other methods such as incorporation into the cheese coatings. It is used at levels ranging from 300 to 2000 parts per million. It is also termed pimaricin.

**Natural Sugar**—*See Turbinado Sugar*.

**Nerol**—A flavoring agent that is a colorless liquid with an odor resembling fresh, sweet roses and contains geranoils and other terpenic alcohols. It is miscible in alcohol, chloroform, and ether; insoluble in water. It is obtained by synthesis. It is also termed Cis-3—,7-dimethyl-2,6-octadien-1-OI.

**Niacin**—A water-soluble B-complex vitamin that is necessary for the growth and health of tissues. It prevents pellagra. It has a solubility of 1 g in 60 ml of water and is readily soluble in boiling water. It is relatively stable in storage and no loss occurs in ordinary cooking. Sources include liver, peas, and fish. It was originally termed nicotinic acid and also functions as a nutrient and dietary supplement.

**Niacinamide**—A nutrient and dietary supplement that is an available form of niacin. Nicotinic acid is pyridine beta-carboxylic acid and nicotinamide, which is another term for niacinamide, is the corresponding amide. It is a powder of good water solubility, having a solubility of 1 g in 1 ml of water. Unlike niacin, it has a bitter taste; the taste is masked in the encapsulated form. Used in fortification of cereals, snack foods, and powdered beverages.

**Nicotinamide**—*See Niacinamide*.

**Nicotinic Acid**—*See Niacin*.

**Nisin**—An antimicrobial agent derived from pure culture fermentations of certain strains of *Streptococcus lactis* Lancefield Group N. Nisin preparation contains nisin, a group of related peptides with antibiotic activity. It is used to inhibit the outgrowth of *Clostridium botulinum* spores and toxin formation in pasteurized cheese spreads and pasteurized process cheese spreads; pasteurized cheese spread with fruits, vegetables, or meats; and pasteurized process cheese spread with fruits, vegetables, or meats.

**Nitrate**—The salt of nitric acid. It is used in meat curing to develop and stabilize the pink color associated with cured meat. By itself, it is not effective in producing the curing reaction until it is chemically reduced to nitrite. It has an effect on flavor and also functions as an antioxidant. It is available as sodium and potassium nitrate, with the sodium form being more common.

**Nitrite**—The salt of nitrous acid. It is used in meat curing to develop and stabilize the pink color associated with a cured meat and to affect flavor and function as an antioxidant. Nitrites convert to nitric oxide, which reacts with the myoglobin pigments (purple-red) to form nitrosomyoglobin (dark red). Nitrosomyoglobin plus heating to 130 to 140°F results in the formation of the stable pigment nitrosohemochrome, resulting in the cured meat color. It has bacteriostatic properties as an inhibitor of *Clostridia* organisms, especially *Clostridium botulinum*, and, therefore, nonsterile canned hams can be produced. Sources are sodium and potassium nitrite, with the sodium form being more commonly used.

**Nitrous Oxide**—A noncombustible gas used as a propellant in certain dairy and vegetable fat whipped toppings contained in pressurized containers.

**(Gamma)-Nonalactone**—A synthetic flavoring agent that is a colorless to yellow liquid of strong, coconut-like odor. It is soluble in most fixed oils, mineral oil, and propylene glycol. It is stable in acids and unstable in alkali and should be stored in glass, tin, or aluminum containers. It is used in coconut flavors and has application in gelatins, puddings, baked goods, candy, and ice cream at 11 to 55 parts per million. It is also termed aldehyde C-18.

**Nonanal**—A flavoring agent that is a colorless or light yellow liquid, with a strong odor resembling an essence of orange and rose. It is soluble in alcohol, most fixed oils, mineral oil, and propylene glycol, but insoluble in glycerin. It is obtained by chemical synthesis. It is also termed aldehyde C-9 and pelargonic aldehyde.

**Nonfat Dry Milk—*See Milk Solids–Not-Fat*.**

**Norbixin—*See Annatto*.**

**Nordihydroguaiaretic Acid—**(NDGA) An antioxidant that has poor solubility and shows evidence of discoloration in the presence of metal salts. It is used to a limited extent to retard rancidity.

**Nutmeg—**A spice obtained from the nutmeg tree *Myristica fragrans*. It is related to mace, which is obtained from the covering of nutmeg. Nutmeg is used in eggnog, cakes, fruit, and puddings.

# O

**Oat**—A grain that is a source of oat flour. It is used in porridge, grits, and oatmeal.

**Oat Flour**—Fine-mesh ground oats with the hull removed. It has some antioxidant properties and is blended with other flours to retard rancidity.

**Oatmeal**—The food produced by grinding oats after removal of the husk.

**(Gamma)-Octalactone**—A synthetic flavoring agent that is a stable, colorless to slightly yellow liquid of peach odor. It should be stored in glass or tin containers. It is used in flavors for peach with applications in baked goods, candy, and ice cream at 5 to 17 parts per million.

**Octanoic Acid**—*See Caprylic Acid.*

**1-Octanol**—A synthetic flavoring agent that is a colorless, stable liquid of sharp fatty odor. It is soluble in alcohol, most fixed oils, mineral oil, and propylene glycol. It should be stored in glass or tin containers. It is used in essential oils for application in beverages, candy, and baked goods at 1 to 3 parts per million. It is also available in the natural form, obtained from natural precursors. It is also termed octyl alcohol.

**Octyl Acetate**—A flavoring agent that is a colorless liquid with a fruity odor resembling orange and jasmine. It is miscible in alcohol, oils, and other organic solvents, and insoluble in water. It is obtained by chemical synthesis.

**Octyl Alcohol**—*See 1-Octanol.*

**Oil of Rue**—A flavoring agent that is the natural substance obtained by steam distillation of the fresh blossoming plants of rue, the perennial herb of several species of *Ruta (Ruta montana* L., *Ruta graveolens* L., *Ruta bracteosa* L., and *Ruta calepensis* L.). It is used in baked foods and baking mixes (10 ppm); frozen dairy desserts and mixes (10 ppm); soft candy (10 ppm); and other food categories (4 ppm).

**Oleic Acid**—An unsaturated fatty acid that functions as a lubricant, binder, and defoamer.

**Oleic Acid Derived from Tall Oil Fatty Acids**—An additive consisting of purified oleic acid separated from refined tall oil fatty acids. It is used in foods as a lubricant, binder, and defoaming agent, and as a component in the manufacture of other food-grade additives. To ensure safe use of the additive, the label should show the common or usual name of the acid, and the words "food grade."

**Oleomargarine**—*See Margarine.*

**Oleoresin Paprika**—A seasoning and colorant that is the solvent-free extraction containing the volatile and nonvolatile flavor components of paprika. It is the closest replacement for paprika. As a colorant, the pigment is a red-orange carotenoid of which the principal carotenoid is capsanthin. It has fair pH and heat stability, and poor light and chemical stability. It is used in sausages, meat products, condiment mixtures, and salad dressings.

**Oleoresins**—Solvent-free extractions from spices that contain the volatile and nonvolatile flavor components. They are the closest replacements for a spice, and are used in seasonings for a variety of foods.

**Olestra**—A fat replacer (sucrose polyester) manufactured using vegetable oil and sucrose to produce a product that is not absorbed or metabolized, passing undigested through the digestive tract. Fat soluble vitamins A, D, E, and K can be carried out of the body with olestra. It is noncaloric because it is not hydrolyzed by digestive lipases. Approved for use in savory snack foods; advisory labeling is required. It is used in fat-free chips and crackers.

**Olive Oil**—The oil obtained from the fruit of olive trees, *Olea europaea*. It is used mainly for salad and cooking oils.

**Onion**—A flavorant, the vegetable *Allium cepa* L., commercially processed into powder, salt, minced, and toasted forms. It is used in meats, sauces, soups, and dips.

**Orange Oil, Bitter**—A flavoring agent that is a yellow-brown liquid with an aromatic odor resembling Seville orange, and an aromatic and bitter taste. Its substance is degraded by light, and its alcohol solutions are neutral to litmus. It is miscible in absolute alcohol and glacial acetic acid, soluble in fixed oils and mineral oil, slightly soluble in propylene glycol, and insoluble in glycerin. It is obtained by cold expression of fresh peel of the fruit of *Citrus aurantium* L. of the *Rutaceae* family.

**Oregano**—A spice made from the dried leaves of *Lippia graveolens*, a perennial of the mint family. There are two strains available. One strain, common to the Mediterranean region, is delicate in fragrance and taste and the other, which is common to Mexico, is quite pungent. It is used in sauces, soups, and pizza.

**Orthophosphates**—Salts of phosphoric acid containing one phosphorous atom. They are made by partially or fully neutralizing phosphoric acid with an alkali. Monobasic orthophosphates have one hydrogen atom replaced with the alkali metal, dibasic have two replaced, and tribasic have three replaced. Examples of orthophosphates include tricalcium phosphate, dipotassium phosphate, disodium phosphate, and trisodium phosphate. Heating under controlled conditions forms condensed phosphates or polyphosphates. Functions include buffering, sequestering, and chelating.

**Orthophosphoric Acid**—*See Phosphoric Acid*.

**Ox Bile Extract**—A yellowish-green soft solid with a part-sweet, part-bitter, disagreeable taste. It is the purified portion of the bile of an ox obtained by evaporating the alcohol extract of concentrated bile. The ingredient is used as a surfactant in food, a surfactant also known as purified oxgall or sodium choleate.

**Oxidized Cornstarch**—Starch produced by treating an aqueous starch suspension with dilute sodium hypochlorite containing a small excess of caustic soda until the desired degree of oxidation is reached. The slurry is then treated with an antichlor, such as sodium bisulfate, adjusted to the desired pH, filtered, washed, and dried. It still retains its original granule structure and is insoluble in water. It is extremely white, has decreased viscosity, is relatively clear, and shows a reduced tendency to thicken when cooled. Its food applications are those where high solids and low viscosity are desired.

**Oxidizing and Reducing Agents**—Substances which chemically oxidize or reduce another food ingredient, thereby producing a more stable product.

**Oxystearin**—A crystallization inhibitor and release agent that is a modified fatty acid composed of the glycerides of partially oxidized stearic and other fatty acids. It is used in vegetable oils to prevent them from clouding in the refrigerator and in griddling fats and oils to prevent food from sticking to the frying pan.

# P

**p-Anisaldehyde**—*See p-Methoxybenzaldehyde.*

**p-Methoxybenzaldehyde**—A flavoring agent that is a colorless or faintly yellow liquid, hawthorn-like odor. It is miscible in alcohol, ether, and most fixed oils, soluble in propylene glycol, insoluble in glycerin, water, and mineral oil. It is obtained by synthesis. It is also termed anise aldehyde and p-anisaldehyde.

**Palmitic Acid**—A fatty acid which is a mixture of solid organic acids from fats consisting principally of palmitic acid with varying amounts of stearic acid. It functions as a lubricant, binder, and defoaming agent.

**Palm Kernel Oil**—An oil obtained from palm kernels. It consists mainly of lauric, myristic, and oleic fatty acids. It resembles coconut oil and is used interchangeably with coconut oil. It is a possible source of stearine, which is a substitute for cocoa butter. It is used in margarine and confectionery.

**Palm Oil**—The oil obtained from the fruit of the palm tree. It has a narrower plastic range than lard and most shortenings which is a disadvantage in shortening applications. It can be used in mixtures with only a moderately adverse effect on the plastic range. It consists mainly of palmitic, oleic, and linoleic fatty acids. It is used in margarine and shortenings.

**Pantothenic Acid**—Vitamin $B_5$, which is a water-soluble vitamin. It is required for proper growth and maintenance of the body and is involved in body processes such as energy release from carbohydrates and metabolism of fatty acids. It is relatively stable through storage and is found in liver, eggs, and meat.

**Papain**—A tenderizer that is a protein-digesting enzyme obtained from the papaya fruit. The enzyme, used in a patented process, is injected into the circulatory system of the live animal and is activated by the heat of cooking to break down the protein, thus tenderizing the beef. The enzyme is inactivated by stomach acids.

**Paprika**—A spice and colorant made from the ground, dried, ripe fruit of the herb *Capsicum annuum* L. It contributes flavor and color to foods. The pod provides red color and has good tinctorial strength, good pH stability, and poor stability to light and oxidation. It is used in meat, fish, sauces, and salad dressings. It is also termed sweet pepper or pimiento. ***See Oleoresin Paprika***.

**Parabens**—Antimicrobial agents that are esters of para-hydroxybenzoic acid. The most common esters are methyl p-hydroxybenzoate and propyl p-hydroxybenzoate. The ethyl and butyl esters have some applications. It is related to benzoic acid and sodium benzoate but is effective over a wide pH range. The parabens are most active against yeasts and molds and are stable to high temperature. They are a white free-flowing powder of fair water solubility at room temperature which improves if the water is heated to 70°C. Methyl paraben is more soluble (0.25 g per 100 ml of water at 25°C) but less effective in mold inhibition than propyl paraben (0.04 g per 100 ml of water at 25°C). It is used in meat and poultry products.

**Para-Hydroxybenzoic Acid**—*See Parabens*.

**Parboiled Rice**—The rice that results from the process of soaking rice in water, draining, pressure cooking to completely gelatinize the starch, drying, and milling. The parboiling process aids the development of stability toward cooking and heat processing. It is used in canned rice products such as soups, casseroles, meat, and rice dinners, such as Spanish rice. The milling of parboiled rice produces parboiled bran.

**Parsley**—A spice made from the dried leaves of *Petroselinum hortense*, of bright green color. It has a high content of vitamins A and C and also contains iron, iodine, copper, and manganese. It is used for garnishing and seasoning, with application in sauces, salads, and soups.

**Partially Hydrogenated Coconut Oil**—*See Coconut Oil*.

**Partially Hydrogenated Oil**—Oil that has been partially hydrogenated (chemical addition of oxygen) to modify the texture from liquid to semi-solid. This conversion raises the melting point. It is used in farinaceous foods, confectionery, and desserts.

**Partially Hydrolyzed Guar Gum**—A source of soluble dietary fiber extracted from guar gum. It does not affect taste or viscosity. It is soluble in cold water with minimal viscosity. It can function to replace fat, increase moisture retention, and stabilize. Uses include cereal, soups, confections, and baked goods.

**Pastry Flour**—A flour obtained from soft wheat. Either straight or clear flour grades may be used because color is not an essential requirement. It is used in white sauces and pastry.

**Patent Flour**—Flour made from the separation of 40 to 90 percent of that portion of the grain that can be milled from a wheat blend. There are various streams to include long patent, medium patent, short patent, first patent, and fancy patent flours.

**Peanut Oil**—The oil obtained from peanuts, consisting principally of the unsaturated fatty acids oleic and linoleic. It is liquid at room temperature, has a specific gravity at 38°C of approximately 1.89 to 0.90, and an iodine number of 85 to 95. It is removed from the nuts by one of two processes, namely, the expeller method, in which the shelled peanuts are cooked with steam, and fed into an expeller press which physically presses the oil from the meal; or the pre-press solvent system, which is comparable to the expeller method except that less pressure is applied, which leaves more oil in the meal, and the remaining meal is solvent-washed, usually with hexane, to dissolve the oil from the meal. The obtained crude oil is refined. The major use of peanut oil is in cooking oils and salad oils. Peanut oil is used in deep-fat frying because of its long frying life and high smoke point. In salad oil, it contributes to the suspension of solids. Other applications include shortening ingredient for doughnuts and cakes.

**Pearl Starch**—*See Cornstarch.*

**Pectic Acid**—Those pectic substances that are essentially void of methoxyl groups and have carboxyl groups only. They have varying degrees of neutralization. The divalent salts are slightly soluble in water and must be converted to the sodium or potassium forms for dissolution. It gels in the presence of calcium or other divalent cations.

**Pectin**—A gum that is the methylated ester of polygacturonic acid. It is obtained from citrus peels and apple pomace. The degree of methylation (DM) or esterification (DE) refers to the percentage of acid groups which are present as the methyl ester. Pectin is divided into two main groups: high methoxy (HM) pectin, having 50 percent or greater esterification, and low methoxy (LM) pectin, having less than 50 percent esterification. These pectins gel under different conditions. The LM pectins are subdivided into low methoxy amidated pectin and low methoxy conventional pectin. *See Pectin, High Methoxy; Pectin, Low Methoxy; Amidated Pectin*.

**Pectin, High Methoxy**—A pectin with a degree of esterification of 50 percent or greater. These pectins gel under acid conditions (pH 3.5 or lower) and high soluble solids (55 percent or higher). The resulting gel sets at varying temperatures into a rigid gel that is not thermally reversible. Applications include jams, jellies, preserves, and bakery fillings. *See Pectin*.

**Pectinic Acid**—A broad group of pectic substances that contain more than a negligible proportion of methyl ester groups and all the unesterified carboxyl groups are free. The divalent salts of pectinic acid are only slightly soluble in water and must be converted to the sodium or potassium form for dissolution.

**Pectin, Low Methoxy**—A pectin with a degree of esterification of less than 50 percent. These pectins gel with divalent ions, such as calcium, over a broad range of pH and soluble solids. The resulting gel is spreadable with some gel structure and sets at varying temperatures. The gel is thermoreversible. Applications include low calorie jams, jellies, and preserves, tomato-based sauces, and low pH milk beverages. *See Pectin*.

**Pelargonic Aldehyde**—*See Nonanal*.

**2-Pentanone**—A flavoring agent that is a clear liquid, colorless, with flowery odor. It is miscible in alcohol and ether and soluble in water. It is obtained by chemical synthesis. It is also termed methyl propyle ketone.

**Pentasodium Tripolyphosphate**—*See Sodium Tripolyphosphate*.

**Pepper**—A spice made from a berry from the vine *Piper nigrum* L. which produces black and white pepper. Black pepper is picked slightly underripe and dried, during which time the characteristic black, wrinkled appearance is attained. White pepper is picked fully ripe and dried, after which the outer hull is removed by attrition to expose the white core. It is used in meat, vegetables, soups, and salads.

**Pepper, Cayenne**—A spice that is not related to the true pepper vine but to the paprika, bell peppers of the *Capsicum* family. It is hot and fiery and used in spreads, dips, and chili sauce.

**Pepper, Red**—The pod of the genus *Capsicum*, variety *C. annuum* L. and *C. frutescens* L. It has a hot, pungent flavor. It is used in barbecue sauce, spicy sauces, and chili powder.

**Peptone**—A polypeptide used as a beer stabilizer.

**Petrolatum**—A release agent, lubricant, and defoaming agent that is a purified mixture of semisolid hydrocarbons obtained from petroleum. It varies in color from white to yellow. It is used in bakery products, dehydrated fruits and vegetables, and egg white solids.

**Petroleum Wax**—A wax used as a masticatory substance in chewing gum base. It is also used as a protective coating on raw fruits and vegetables and as a fruit defoamer.

**Phenethyl Phenylacetate**—A flavoring agent that is a colorless or pale yellow liquid, with an odor resembling roses and hyacinth, which becomes solid at <26°C (78.8°F). It is soluble in alcohol, insoluble in water. It is obtained by chemical synthesis.

**Phenylacetic Acid**—A flavoring agent that is crystalline (white, glistening), with unpleasant, persisting odor resembling geranium leaf and rose when diluted. It is soluble in most fixed oils and glycerin, slightly soluble in water, and insoluble in mineral oil. It is obtained by chemical synthesis. It is also termed A-toluic acid.

**Phenylethyl Anthranilate**—A synthetic flavoring agent that is a stable, white to yellow crystal of grape and orange blossom odor. It should be stored in glass or polyethylene-lined containers. It is used for flavors such as grape and cherry in applications such as beverages, ice cream, candy, and baked goods at 2 to 6 parts per million.

**Phenylethyl Isobutyrate**—A synthetic flavoring agent that is a stable, colorless to light yellow liquid of fruity odor of floral note. It is soluble in alcohol and practically insoluble in water. It should be stored in glass or tin-lined containers. It is used in flavors for peach with applications in beverages, ice cream, candy, and baked goods at 3 to 13 parts per million.

**Phosphate**—Any salt of phosphoric acid. The salts include disodium phosphate, trisodium phosphate, sodium hexamethaphosphate, and others. They play a variety of roles such as sequestrants, emulsifiers, solubility enhancers, and buffers in a variety of foods.

**Phosphated Flour**—Flour to which monocalcium phosphate is added at not less than 0.25 percent and not more than 0.75 percent. It is used in baked goods.

**Phosphoric Acid**—An acidulant that is an inorganic acid produced by burning phosphorus in an excess of air, producing phosphorus pentoxide which is dissolved in water to form orthophosphoric acid of varying concentrations. It is a strong acid which is soluble in water. The acid salts are termed phosphates. It is used as a flavoring acid in cola and root beer beverages to provide desirable acidity and sourness. It is used as a synergistic antioxidant in vegetable shortenings. In yeast manufacture, it is used to maintain the acidic pH and provide a source for phosphorus. It also functions as an acidulant in cheese. It is also termed orthophosphoric acid.

**Pimaricin**—*See Natamycin*.

**Pimiento**—*See Paprika*.

**Piperonyl Acetate**—A synthetic flavoring agent that is a stable, colorless to light yellow liquid of heliotrope odor. It should be stored in glass or resin-lined containers. It is used in flavors for berry notes with applications in beverages, candy, ice cream, and baked goods at 50 to 90 parts per million.

**Piperonyl Isobutyrate**—A synthetic flavoring agent that is a moderately stable, colorless to light yellow liquid of fruity odor. It should be stored in glass or resin-lined containers. It is used in flavors for cherry, berry, and peach aroma with applications in beverages, candy, and baked goods at 1 to 4 parts per million.

**Plasticizer**—*See Softener*.

**Polydextrose**—A bulking agent that is a randomly bonded condensation polymer of dextrose containing small amounts of bound sorbitol and citric acid. It is a water-soluble powder providing a pH range of 2.5 to 3.5. It is partially metabolized which results in a caloric value of one calorie per gram. As a reduced-calorie bulking agent, it can partially replace sugars and in some cases fats in reduced-calorie foods. It also functions as a bodying agent and humectant. Applications include desserts, specific baked goods, frozen dairy desserts, chewing gum, and candy. Usage levels vary according to application, but examples are frozen dessert, 13 to 14 percent; puddings, 8 to 9 percent; and cake, 15 to 16 percent.

**Polyethylene Glycol**—A binder, coating agent, dispersing agent, flavoring adjuvant, and plasticizing agent that is a clear, colorless, viscous, hygroscopic liquid resembling paraffin (white, waxy, or flakes), with a pH of 4.0 to 7.5 in 1:20 concentration. It is soluble in water (MW 1000) and many organic solvents.

**Polyglycerate 60**—*See Ethoxylated Mono- and Diglycerides*.

**Polyglycerol Esters of Fatty Acids**—Emulsifiers that are mixed partial esters formed by reacting polymerized glycerols with edible fats, oils, or fatty acids. They vary in degree of polymerization, and by varying the proportions and fats to be reacted, a diverse class of products is obtainable. The esters range from hydrophilic to lipophilic. They are used in cake mixes for volume and texture, in confectionery for gloss, in whipped toppings for aeration, and in flavors and colors as a solubilizer. Typical usage range is from 0.1 to 1.0 percent.

**Polyoxyethylene (20) Mono- and Diglycerides of Fatty Acids**—*See Ethoxylated Mono- and Diglycerides*.

**Polyoxyethylene Sorbitan Ester**—*See Polyoxyethylene Sorbitan Fatty Acid Esters*.

**Polyoxyethylene Sorbitan Fatty Acid Esters**—Emulsifiers made by reacting ethylene oxide with sorbitan esters to increase their hydrophilic properties. They are generally used in oil and water emulsions in combination with lipophilic emulsifiers such as mono- and diglycerides or sorbitan monostearates to produce a wide variety of effects. They are also termed polysorbates, which include polysorbate 80 (polyoxyethylene [20] sorbitan monooleate), polysorbate 60 (polyoxyethylene [20] sorbitan monostearate), and polysorbate 65 (polyoxyethylene [20] sorbitan tristearate). They can solubilize essential and vitamin oils. They are used in panned coatings to reduce panning time, in coffee whiteners to prevent oiling-off, and in ice cream to produce dryness and overrun. Typical usage level ranges from 0.05 to 0.10 percent.

**Polyoxyethylene (20) Sorbitan Monooleate**—An emulsifier produced by reacting oleic acid with sorbitol to yield a product which is reacted with ethylene oxide. It is a nonionic, water-dispersible surface-active agent that is very soluble in water. It is also termed polysorbate 80. It is used in ice cream and frozen desserts for overrun and dryness; as a disperser of flavors and colors in pickles; and for volume and texture in baked goods. It is frequently used with mono- and diglycerides at usage levels ranging from 0.05 to 0.10 percent.

**Polyoxyethylene (20) Sorbitan Monostearate**—An emulsifier manufactured by reacting stearic acid with sorbitol to yield a product which is reacted with ethylene oxide. It is a nonionic, water-dispersible surface-active agent which is very hydrophilic. It is also termed polysorbate 60. It is used in whipped vegetable toppings for overrun and lightness; in cakes for increased volume and fine grain; in icings and confectionery for lightness and syneresis control; and in salad dressing for emulsion stability. It is frequently used with sorbitan monostearate or mono- and diglycerides. The typical usage range is 0.10 to 0.40 percent.

**Polyoxyethylene (20) Sorbitan Tristearate**—An emulsifier manufactured by reacting stearic acid with sorbitol to yield a product which is then reacted with ethylene oxide. It is a nonionic surface-active agent which is dispersible in fat, oil, and water. It is also termed polysorbate 65. It is used in frozen desserts, cakes, and coffee whiteners. It is frequently used with sorbitan monostearates or mono- and diglycerides. Typical usage range is 0.10 to 0.40 percent.

**Polyoxyl (40) Stearate**—An emulsifier and antifoaming agent used in processed foods, fruit jellies, and sauces.

**Polyphosphates**—Phosphates containing two or more phosphorous atoms per molecule, being formed when orthophosphates are heated under controlled conditions. Pyrophosphates have two phosphorous atoms (for example, sodium acid pyrophosphate); tripolyphosphates having three phosphorous atoms (for example, sodium tripolyphosphate). Further heating polyphosphates and chilling forms a longer chain (for example, sodium hexametaphosphate). Functions include sequestering, buffering, and chelating. They are also termed condensed phosphates.

**Polysorbate 60**—*See Polyoxyethylene (20) Sorbitan Monostearate*.

**Polysorbate 65**—*See Polyoxyethylene (20) Sorbitan Tristearate*.

**Polysorbate 80**—*See Polyoxyethylene (20) Sorbitan Monooleate*.

**Polysorbates**—*See Polyoxyethylene Sorbitan Fatty Acid Eaters*.

**Pomace**—Ground apple or fleshy fruit in the dry form.

**Popcorn**—Indian corn that explodes when exposed to dry heat due to the expansion of the kernel.

**Poppy Seed**—A seasoning that is a seed of *Papaver somniferum* L. Poppy seeds have a nutty flavor. They are used in breads, cakes, and butter sauce for vegetables, lending a nutlike flavor.

**Potassium Acid Tartrate—*See Cream of Tartar*.**

**Potassium Alginate**—A gum that is the potassium salt of alginic acid. It is soluble in cold water, forming a viscous colloidal solution. It functions as a stabilizer, thickener, and gelling agent. It is used in dietetic foods, low-sodium foods, dry mixes, and dental impression material. Typical usage levels range from 0.05 to 0.50 percent.

**Potassium Bicarbonate**—An alkali and leavening agent obtained as colorless prisms or white powder. It is very soluble, with 1 g dissolving in 2.8 ml of water. Upon heating, it liberates carbon dioxide which provides leavening in baked goods. It is also used in confectionary products.

**Potassium Bisulfite**—A preservative that retards bacterial action, prevents discoloration, and functions as an antioxidant. It is not used in meats or in food recognized as a source of vitamin $B_1$, and it is not used on fruits or vegetables intended to be served raw or presented as fresh.

**Potassium Bitartrate—*See Cream of Tartar*.**

**Potassium Bromate**—A dough conditioner that exists as white crystals or powder and is soluble in water. It exists in the anhydrous form as white granular powder and in the hydrated form as small white crystals or granules. It is used to age and improve the baking properties of flour. It is used with potassium iodate and azodicarbonamide to modify the protein in bread flour to promote the desired properties of loaf volume and shape. It is used in baked goods.

**Potassium Carbonate**—A general purpose food additive and alkali. It is hygroscopic and the aqueous solutions are strongly alkaline. It has a solubility of 1 g in 1 ml of water at 25°C. It is used as a flavoring agent and processing aid, and to control pH. It is used in soups to neutralize acidity.

**Potassium Carrageenan—*See Carrageenan*.**

**Potassium Caseinate—*See Caseinates*.**

**Potassium Chloride**—A nutrient, dietary supplement, and gelling agent that exists as crystals or powder. It has a solubility of 1 g in 2.8 ml of water at 25°C and 1 g in 1.8 ml boiling water. Hydrochloric acid, and sodium chloride and magnesium chloride diminish its solubility in water. It is used as a salt substitute and mineral supplement. It has optional use in artificially sweetened jelly and preserves. It is used as a potassium source for certain types of carrageenan gels. It is used to replace sodium chloride in low-sodium foods.

**Potassium Citrate, Monohydrate**—A sequestrant and buffer that exists as crystals or powder. It is slightly hygroscopic and possesses the advantageous properties of citric acid without having its acid reaction. A 1 percent solution has a pH of 7.5 to 9.0. It reacts with metal ions such as calcium, magnesium, and iron to form a complex. It is soluble in water with a solubility of 1.8 g in 1 ml 20°C water and 2 g in 1 ml of 80°C water. It is found in artificially sweetened jelly and in certain milk and meat products. Uses include processed cheese, puddings, and dietetic foods in which sodium is undesirable. It is also termed tripotassium citrate.

**Potassium Dihydrogen Orthophosphate**—*See Monopotassium Phosphate*.

**Potassium Gluconate**—A nutritional source of potassium used in fortification. It has solubility in water at 20°C of greater than 900 grams per liter, and a pH of approximately 7.0 at 1 percent solution. It can be used as a partial replacement for sodium chloride to reduce the level of sodium, such as in cheese and bakery goods.

**Potassium Hydrogen Tartrate**—*See Cream of Tartar*.

**Potassium Hydroxide**—A water-soluble food additive and bleaching agent. Upon exposure to air it readily absorbs carbon dioxide and moisture and deliquesces. It is used to destroy the bitter chemical constituents in olives that will be used as black olives.

**Potassium Iodate**—A source of iodine made by reacting iodine with potassium hydroxide. It is a crystalline powder which is more stable than iodide. It has a solubility of 1 g in 15 ml of water. It is used as a fast-acting dough improver; it is used with potassium bromate as an oxidizing agent to modify the protein in bread flour which promotes loaf volume and shape. It is used in baked goods.

**Potassium Iodide**—A source of iodine and a nutrient and dietary supplement. It exists as crystals or powder and has a solubility of 1 g in 0.7 ml of water at 25°C. It is included in table salt for the prevention of goiter.

**Potassium Lactate**—A flavor enhancer that is the potassium salt of lactic acid. It is a hygroscopic, white, odorless solid and is prepared commercially by the neutralization of lactic acid with potassium hydroxide. It is used as a flavoring agent and enhancer in some meat and poultry products, a humectant, and a pH control agent.

**Potassium Metabisulfite**—A chemical preservative and antioxidant obtained as white or colorless crystals, powder, or granules. It is soluble in water and insoluble in alcohol. The sulfite salt yields sulfurous acid at a low pH. It is used as a food preservative.

**Potassium Metaphosphate**—A substitute for sodium phosphate in low-sodium foods. It also functions as a fermentation nutrient and buffer.

**Potassium Nitrate**—A preservative and color fixative in meats which exists as colorless prisms or white granules or powder. It has a solubility of 1 g in 3 ml of water at 25°C. *See Nitrate.*

**Potassium Nitrite**—A color fixative in meats which exists as white or yellowish granules or cylindrical sticks. It is very soluble in water. *See Nitrite.*

**Potassium Oleate**—The potassium salt of oleic acid. It is used as a binder, emulsifier, and anticaking agent.

**Potassium Palmitate**—The potassium salt of palmitic acid. It is used as a binder, emulsifier, and anticaking agent.

**Potassium Phosphate Dibasic**—*See Dipotassium Phosphate.*

**Potassium Phosphate Monobasic**—*See Monopotassium Phosphate.*

**Potassium Sodium Tartrate**—*See Sodium Potassium Tartrate.*

**Potassium Sorbate**—A preservative that is the potassium salt of sorbic acid. It is a white crystalline powder which is very soluble in water, with a solubility of 139 g in 100 ml at 20°C. This solubility allows for solutions of high concentration which can be used for dipping and spraying. It is effective up to pH 6.5. It has approximately 74 percent of the activity of sorbic acid, therefore requiring higher concentrations to obtain comparable results as sorbic acid. It is effective against yeasts and molds and is used in cheese, bread, beverages, margarine, and dry sausage. Typical usage levels are 0.025 to 0.10 percent.

**Potassium Stearate**—The potassium salt of stearic acid. It is used as a binder, emulsifier, anticaking agent, and as a placticizer in chewing gum base.

**Potassium Sulfate**—A flavoring agent that occurs naturally, consisting of colorless or white crystals or crystalline powder having a bitter, saline taste. It is prepared by the neutralization of sulfuric acid with potassium hydroxide or potassium carbonate.

**Potato Starch**—A starch obtained from potatoes. It provides long body and clarity to food. It is used mainly in those countries in which it is the principal commercial starch. Applications include Danish desserts, soups, and gravies.

**Powdered Sugar**—A sweetener obtained by pulverizing granulated sugar and adding approximately 3 percent cornstarch. The blend is ground to the desired fineness, that is, 4X, 6X, or 8X. It is very soluble in water. Applications include confectioneries and icings.

**Precipitated Calcium Phosphate**—*See Tricalcium Phosphate*.

**Pregelatinized Starch**—Starch that has been processed to permit swelling in cold water, unlike natural starch which requires heating. The processing usually consists of cooking starch slurries, drying, and grinding to a fine powder. It is used in instant puddings, cake mixes, and soup mixes at 1 to 5 percent. It is also termed gelatinized wheat starch.

**Preservatives**—Antimicrobial agents used to preserve food by preventing growth of microorganisms and subsequent spoilage, including fungicides, mold, and rope inhibitors. The preservatives most widely used are the benzoates (sodium benzoate), sorbates (sorbic acid and potassium sorbate), and the propionates (sodium or calcium propionate), which are organic acids or their salts. Acidulants are used as preservatives because they increase the acidity of food, which can reduce growth of bacteria. Acidulants used include acetic acid, adipic acid, citric acid, fumaric acid, lactic acid, and phosphoric acid.

**Processing Aids**—Substances used as manufacturing aids to enhance the appeal or utility of a food or food component, including clarifying agents, clouding agents, catalysts, flocculents, filter aids, and crystallization inhibitors.

**Propane**—An aerating agent used in combination with chloropentafluoroethane or octafluorocyclobutane as a propellant and aerating agent for foamed or sprayed foods.

**Propellants, Aerating Agents, and Gases**—Gases used to supply force to expel a product or used to reduce the amount of oxygen in contact with the food in packaging.

**Propionic Acid**—The acid source of the propionates. Propionic acid in the liquid form has a strong odor and is corrosive, so it is used as the sodium, calcium, and potassium salts as a preservative. These yield the free acid in the pH range of the food in which they are used. It functions principally against mold. *See Calcium Propionate; Sodium Propionate*.

**Propylene Glycol**—A humectant and flavor solvent that is a polyhydric alcohol (polyol). It is a clear, viscous liquid with complete solubility in water at 20°C and good oil solvency. It functions as a humectant, as do glycerol and sorbitol, in maintaining the desired moisture content and texture in foods such as shredded coconut and icings. It functions as a solvent for flavors and colors that are insoluble in water. It is also used in beverages and candy.

**Propylene Glycol Alginate**—A gum that is the propylene glycol ester of alginic acid, which is obtained from kelp. As compared to sodium alginates, it has reduced sensitivity to acid and calcium salts. It functions in acidic systems. It functions as a thickener, stabilizer, and emulsifier in beer, salad dressings, syrups, and fruit drinks.

**Propylene Glycol Ester**—*See Propylene Glycol Mono- and Di-Esters*.

**Propylene Glycol Mono- and Di-Esters**—A lipophilic emulsifier that consists of propylene glycol esters of fatty acids, such as palmitic and stearic. It is used to increase the whipping ability and aeration in cake batters and whipped toppings.

**Propylene Glycol Monostearate**—A lipophilic emulsifier that is a propylene glycol ester. It is used as a dispersing aid in nondairy creamers; as a crystal stabilizer in cake shortenings and whipped toppings; and as an aeration increaser in cake batters, icings, and toppings. It is also used in oils and shortenings.

**Propyl 2-Furanacrylate**—A synthetic flavoring agent that is a stable, colorless to light yellow liquid of fruity odor. It should be stored in glass or tin-lined containers. It is used in flavors for apple, pear, and raspberry with applications in beverages, candy, and baked goods at 1 to 3 parts per million.

**Propyl Gallate**—An antioxidant that is the *n*-propylester of 3,4,5-trihydroxybenzoic acid. Natural occurrence of propyl gallate has not been reported. It is commercially prepared by esterification of gallic acid with propyl alcohol followed by distillation to remove excess alcohol.

**Propyl Heptanoate**—A synthetic flavoring agent that is a stable, colorless liquid of fruity odor. It should be stored in glass, tin, or resin-lined containers. It is used in apple flavors and modified coffee. It has applications in beverages, ice cream, candy, and baked goods at 4 to 18 parts per million.

**Propyl p-Hydroxybenzoate**—*See Parabens*.

**Propyl Paraben**—*See Parabens*.

**Psyllium**—A gum obtained from the plant of the *Plantago* genus. It hydrates slowly to form a viscous dispersion of concentrations up to 1 percent. A clear, gelatinous mass is formed at 2 percent. It is used in bulk laxatives.

**Pyridoxine**—Vitamin $B_6$, a water-soluble vitamin with a solubility of 1 g in 5 ml of water. It functions in the utilization of protein and is an essential nutrient in enzyme reactions. It is necessary for proper growth. During processing, there is a loss due to leaching of the vitamin in water. It is destroyed by high temperatures, high irradiation, and exposure to light. During storage, loss increases with temperature and storage time. It is found in liver, eggs, and meats. *See Pyridoxine Hydrochloride*.

**Pyridoxine Hydrochloride**—An acid form of vitamin $B_6$, a water-soluble vitamin. It is soluble in water, and slightly soluble in alcohol. It is slowly affected by sunlight and is reasonably stable in air. It has a pH of 2.3 to 3.5. It is also termed vitamin $B_6$ hydrochloride. *See Pyridoxine*.

# Q

**Quicklime**—*See Calcium Oxide.*

**Quince Seed**—A gum produced from the fruit of the quince tree *Cydonia oblonga*. It hydrates slowly to form a highly viscous dispersion at concentrations up to 1.5 percent. Above 2 percent, a slimy, muscilaginous mass is formed. It is principally used in the cosmetic industry. It is also termed gum quince seed, semen cydonia, golden apple seed, and cydonia seed.

**Quinine**—A flavorant naturally obtained from the cinchona tree. It is used as a bitter flavoring in beverages such as quinine water, tonic water, and bitter lemon. Quinine sulfate and quinine hydrochloride are cleared for use as a flavor in carbonated beverages at levels less than 83 parts per million.

# R

**Raisin**—A dried grape used as a fruit and as an ingredient in cereals, baked goods, and desserts.

**Raisin Seed Oil**—*See Grape Seed Oil*.

**Rapeseed Oil**—The oil derived from seeds of *Brassica campestris* or *B. napsus* of the family *Cruciferae* and related trees. It can function as a stabilizer and thickener in peanut butter and as an emulsifier in cake mix shortenings.

**Rapeseed Oil, Fully Hydrogenated**—A stabilizer and thickener. A mixture of triglycerides in which the fatty acid composition is a mixture of saturated fatty acids. The fatty acids are present in the same proportions which result from the full hydrogenation of fatty acids occurring in natural rapeseed oil. Obtained from the *napus* and *campestris* varieties of *Brassica* of the family *Cruciferae*. Prepared by full hydrogenation of refined and bleached rapeseed oil at 310°F, using a catalyst such as nickel, until the iodine number is 4 or less. Used as a stabilizer and thickener in peanut butter.

**Rapeseed Oil, Low Erucic Acid**—Fully refined, bleached, and deodorized oil obtained from certain varieties of *Brassica napus* or *B. campestris* of the family *Cruciferae*. Chemically, low erucic acid rapeseed oil is a mixture of triglycerides, composed of both saturated and unsaturated fatty acids, with an erucic acid content of no more than 2 percent of the component fatty acids. It may be partially hydrogenated to reduce the proportion of unsaturated fatty acids. Low erucic acid rapeseed oil and partially hydrogenated low erucic acid rapeseed oil are used in food, except in infant formula. It is also termed canola oil.

**Rapeseed Oil, Superglycerinated Fully Hydrogenated**—An emulsifier that is a mixture of mono- and diglycerides with triglycerides as a minor component. The fatty acid composition is a mixture of saturated fatty acids present in the same proportions as those resulting from the full hydrogenation of fatty acids in a natural rapeseed oil. It is made by adding excess glycerol to the fully hydrogenated rapeseed oil and heating, in the presence of a sodium hydroxide catalyst, to 330°F under partial vacuum and steam sparging agitation. The ingredient is used as an emulsifier in shortenings for cake mixes.

**Raw Sugar**—A sweetener that is an intermediate product, containing nonsugar impurities, thus being less refined than white sugar. It is made by crushing and shredding sugar cane to extract the juice which is processed to yield raw sugar and upon further processing yields refined cane sugar.

**Red Durum Wheat**—Wheat obtained from the durum wheat kernel. It is used in macaroni and spaghetti products. *See Durum Wheat.*

**Reduced Lactose Whey**—The portion of milk obtained by the removal of lactose from whey; the lactose content of the finished dry product does not exceed 60 percent. As with whey, reduced lactose whey can be used in fluid, concentrate, or a dry product form. The acidity of reduced lactose whey may be adjusted by the addition of safe and suitable pH-adjusting ingredients.

**Reduced Minerals Whey**—The substance obtained by the removal of a portion of the minerals from whey; the dry product does not contain more than 7 percent ash. As with whey, reduced minerals whey can be used in fluid, concentrate, or a dry product form. The acidity of reduced minerals whey may be adjusted by the addition of safe and suitable pH-adjusting ingredients.

**Reducing Sugar**—A sugar that can chemically react with copper in an alkaline solution. It combines with nitrogen compounds at elevated temperature to produce a browning "Maillard" reaction which contributes to the production of a brown crust in baked goods. It is used in the production of caramel color. Dextrose and fructose are reducing sugars.

**Regular Constarch**—*See Cornstarch.*

**Rennet**—A milk coagulant that is the concentrated extract of rennin enzyme obtained from calves' stomachs (calf rennet) or adult bovine stomachs (bovine rennet). The commercial saline extract of rennin contains a little pepsin, some sodium chloride, and some boric acid, sodium benzoate, or propylene glycol as a preservative. In the paste form, it also contains lipase. In the paste form it is used in Italian-type cheeses. It is used to coagulate milk into curd in making cheese and junket. A microbial rennet and a pepsin rennet also exist. **See Rennin**.

**Rennet Casein**—The product that results from the precipitation of pasteurized milk with a rennet enzyme. Rennet casein requires a pH above 9 to dissolve, as compared to acid casein, which can be dissolved in alkali at a pH as low as 6.5. Rennet casein can be dispersed at lower pH by adding a complex phosphate such as sodium tripolyphosphate. This results in a casein of good emulsifying, whipping, foam stability, and water-binding properties. Uses include imitation cheese.

**Rennin**—A milk coagulant that is an enzyme obtained from the abomasum portion of the stomach of suckling mammals. It is most active at pH 3.8. One part purified rennin will coagulate more than five million parts of milk. The commercial extract of rennin is termed rennet. It is used to coagulate milk in making cheese, junket, and custard. **See Rennet**.

**Retinol**—The fat-soluble vitamin A which is required for new cell growth and prevention of night blindness. There is no appreciable loss by heating or freezing, and it is stable in the absence of air. Sources include liver, fortified margarine, egg, and milk. Vitamin A palmitate can be found in frozen egg substitute.

**Rhodinol**—A flavoring agent that is a colorless liquid, with an odor resembling rose. It is soluble in most fixed oils, mineral oil, and propylene glycol, insoluble in glycerin. It is usually obtained from reunion germanium oil.

**Riboflavin**—The water-soluble vitamin $B_2$, required for healthy skin and the building and maintaining of body tissues. It is a yellow to orange-yellow crystalline powder. It acts as a coenzyme and carrier of hydrogen. It is stable to heat but may dissolve and be lost in cooking water. It is relatively stable to storage. Sources include leafy vegetables, cheese, eggs, and milk.

**Rice Bran Oil**—An oil made from rice bran that consists mainly of oleic, linoleic, and palmitic fatty acids. It is used in salad oil, cooking oil, and hydrogenated shortenings.

**Rice Bran Wax**—A refined wax obtained from rice bran. It is insoluble in water. It is used in candy, fresh fruits, and vegetables as a coating and as a plasticizing material in chewing gum.

**Rice Flour**—The flour made from different varieties of long-, medium-, and short-grain rice, usually obtained from the broken milled rice. The chemical composition is the same as that of the whole rice. The flour does not contain gluten and, as a result, doughs made from it do not retain the gases generated during baking. Rice flours from different varieties display characteristic viscosity patterns during the heating and cooling of their pastes. In general, rice with starch of an amylose content greater than 22 percent has a relatively low peak viscosity and forms a rigid gel on cooling (high set-back viscosity). Rice with a starch low in amylose has a high peak and low set-back viscosity. Rice flour is used in formulated baby foods, breakfast foods, meat products, and breading.

**Rice Starch**—The starch obtained from rice. It forms tender, opaque gels. It has some use in puddings.

**Rochelle Salt**—*See Sodium Potassium Tartrate.*

**Rosemary**—A spice made from the dried leaves of *Rosmarinus officinalis* L., an evergreen shrub. It has a medicinal, menthol flavor. It is available in whole and ground forms. It is used in soups, poultry, and meats, especially lamb.

**Rum Ether**—A synthetic flavoring agent that is a stable, colorless to yellow liquid of ethereal rum-like note. It should be stored in glass and stainless steel containers. It is used to intensify rum flavors for application in beverages, candy, and ice cream at 67 to 320 parts per million and in alcoholic beverages at 1600 parts per million. It is also termed ethyl oxyhydrate.

**Rye**—A cereal crop that is a source of rye flour. It is used as a bread grain.

**Rye Flour**—The flour obtained by milling rye. It is available in white, medium, and dark grades and has a distinct flavor. It is usually diluted with wheat flour in order to make it more palatable. It is used in bread making.

# S

**Saccharin**—A non-nutritive synthetic sweetener which is 300 to 400 times sweeter than sucrose. It is nonhygroscopic and has a bitter aftertaste and a stability problem in cooked, canned, or baked goods. It is slightly soluble in water with a solubility of 10 g in 100 g of water at 25°C, but the solubility improves in boiling water. As sodium saccharin, there are two forms: 1,2-benzisothiazolin-3-one-1,1-dioxide, sodium salt dihydrate, with a solubility of 1 g in 1.2 ml water; and 1,2-benzisothiazolin-3-one-1,1-dioxide, sodium salt. Calcium saccharin (chemical name: 1,2-benzisothiazolin-3-one-1,1-dioxide, calcium salt) is used where low sodium content and reduced after-taste are required. It is used in low-calorie foods such as jam, beverages, and desserts. It is also termed sodium benzosulfimide.

**Sacharose**—*See Sugar.*

**Safflower Oil**—An unsaturated oil obtained from the safflower seed of the plant *Carthamus tinctorius*. It consists mainly of linoleic and oleic fatty acids. It is used principally as a drying oil in the United States.

**Saffron**—A spice obtained from the dried stigmas of the fall-flowering *Crocus sativus* L. The flower stigma is of intense yellow color. It has a powerful, somewhat bitter aroma. It is used in breads, fish, chicken, sauces, and rice dishes.

**Sage**—A spice made from the dried leaves of the shrub *Salvia officinalis* L. It has a strong, fragrant odor. It is available industrially as whole leaf, cut, rubbed, and ground to determined granulations. It is used in pork, soups, poultry seasonings, and fish.

**Sago Starch**—The starch obtained from the sage palm *Metroxylon sagus* or *M. rumphii* and the palm fern *Cycas circinalis*. It forms high-strength gels which lose their clarity upon standing. It is used in confections and puddings.

**Saint John's Bread**—*See Locust Bean Gum.*

**Salt**—A seasoning and preservative whose chemical composition is sodium chloride, about 40 percent sodium and 60 percent chlorine by weight. It contains not less than 97.5 percent sodium chloride after drying, while high-grade salt contains 99.8 percent sodium chloride. Salt production can be by solar evaporation, rock salt mining, and vacuum pan evaporation. The method selected depends on climate, character of the deposit, and type of salt required. Sea salt is obtained from the sea. Seasoned salt contains added flavors. It is available in several particle sizes (coarse, flake, fine) and shapes (flake, cube) which relate to density, solubility, flow, blending, and adherence. It is used as a carrier for dry or semidry ingredients or as an ingredient in prepared mixes. It is used in cheese, butter, and salted nuts for flavor. It is used in cheese manufacture to help remove the whey and suppress the growth of unwanted organisms, in sausage as a seasoning and curing agent, and in baked goods, pickles, and sauerkraut for flavor and fermentation control.

**Santalol**—A flavoring agent that is a colorless or pale yellow liquid, with odor resembling sandalwood. It is soluble in alcohol, fixed oils, mineral oil, and propylene glycol; and insoluble in water and glycerin. It is obtained from a sandalwood oil source.

**Savory**—A spice that is the dried leaves and flowering tops of the plant *Satureia hortenis* L. The two distinct varieties are summer savory and winter savory. Summer savory is generally preferred because it has a more delicate flavor and is less resinous. It is used in soups, salads, and sauces.

**Sea Salt**—*See Salt*.

**Seasoned Salt**—*See Salt*.

**Self-Rising Flour**—White flour to which sodium bicarbonate and one or more of the acid-reacting substances are added, that is, monocalcium phosphate, sodium acid pyrophosphate, or sodium aluminum phosphate. It is seasoned with salt. The inclusion of these ingredients provides a leavening system that allows the flour to rise when wetted in the preparation of baked goods.

**Semen Cydonia**—*See Quince Seeds*.

**Semolina**—The purified ground middlings of durum wheat. It contains bran specks. Durum semolina is ground so that not more than 3 percent passes through a number 100 U.S. sieve. It takes longer to cook and is more resistant to overcooking than flour and results in less cloudiness in the water. It has a 50 percent relative protein efficiency as compared to nonfat dry milk. It is used in macaroni and spaghetti products. It is also termed durum semolina.

**Sequestrants (Chelating Agents)**—Substances which combine with polyvalent metal ions to form a soluble metal complex, to improve the quality and stability of products. Examples include calcium citrate, calcium diacetate, calcium hexametaphosphate, citric acid, dipotassium phosphate, disodium phosphate, isopropyl citrate, monobasic calcium phosphate, monoisopropyl citrate, potassium citrate, sodium acid phosphate, sodium citrate, sodium gluconate, sodium hexametaphosphate, sodium metaphosphate, sodium phosphate, sodium pyrophosphate, sodium tripolyphosphate, stearyl citrate, and tetra sodium pyrophosphate.

**Sesame Oil**—The oil obtained from sesame seeds. It consists principally of oleic and linoleic fatty acids. It has resistance to oxidation. It is used in vegetable shortenings, salad oil, and cooking oil, and is found in frozen chicken chow mein.

**Sesame Seed**—The seed of the plant *Sesamum indicum* L. It has a sweet, "nutty" flavor. It yields sesame oil. It is used in breads, meats, and vegetables. It is also termed benne.

**Shallot**—*Allium ascalonicum*, a member of the onion family. It ranges in size from walnut to small fig and is milder than the onion. It can be substituted for the onion and is used in sauces, dressings, soups, and meats.

**Shortening**—Any animal or vegetable fat or oil that "shortens" or retards the development of gluten strands in baked goods for the purpose of producing a tender, crisp texture. Solid fats are most commonly used instead of oils because of their plastic nature. It is used in baked goods.

**Silica, Amorphous**—*See Silicon Dioxide.*

**Silicon Dioxide**—An anticaking agent, carrier, and dispersant that can absorb approximately 120 percent of its weight and remain free flowing. It is used in salt, flours, and powdered soups to prevent caking caused by moisture. It is also used in powdered coffee whitener, vanilla powder, baking powder, dried egg yolk, and tortilla chips. The usage level ranges from 1 to 2 percent. It is also termed silica, amorphous.

**Skeletal Meat**—The edible part of the animal that is muscle tissue attached to the bone. It includes the shoulder and side of pork, brisket, flank, and round of beef. It is an ingredient in sausage.

**Skim Milk**—Milk from which sufficient fat has been removed to reduce the milkfat content to less than 0.5 percent. It is used in the manufacture of certain cheese varieties, casein, and lactose. It is an ingredient in frozen deserts, baked goods, and confectionery. It is also consumed as a beverage.

**Skim Milk Powder**—*See Milk Solids–Not-Fat.*

**Slaked Lime**—*See Calcium Hydroxide.*

**Smoke Flavoring**—A flavorant that can be obtained in the form of liquid smoke derived from burning hardwoods such as maple and hickory or as synthetic smoke made by synthesis. It is used to impart flavor and aroma to bacon, ham, and sausage.

**Soda Alum**—*See Aluminum Sodium Sulfate.*

**Sodium**—A metal element that performs bodily functions.

**Sodium Acetate**—A source of acetic acid that is obtained as crystals or powder. It has a solubility of 1 g in 0.8 ml of water.

**Sodium Acetate, Anhydrous**—A source of acetic acid obtained as a granular powder. It has a solubility of 1 g in 2 ml of water.

**Sodium Acid Carbonate**—*See Sodium Bicarbonate.*

**Sodium Acid Phosphate**—*See Monosodium Phosphate.*

**Sodium Acid Pyrophosphate**—A leavening agent, preservative, sequestrant, and buffer which is mildly acidic with a pH of 4.1. It is moderately soluble in water, with a solubility of 15 g in 100 ml at 25°C. It is used in doughnuts and biscuits for its variable gas release rate during the mixing, bench action, and baking process. It is used in baking powder as a leavening agent. It is used in canned fish products to reduce the level of undesired struvite crystals (magnesium ammonium phosphate hexahydrate) by complexing the magnesium. It is used to sequester metals in processed potatoes. It is also termed SAPP, disodium dihydrogen pyrophosphate, acid sodium pyrophosphate, disodium diphosphate, and sodium pyrophosphate.

**Sodium Alginate**—A gum obtained as a sodium salt of alginic acid, which is obtained from seaweed. It is cold- and hot-water soluble, producing a range of viscosities. It forms irreversible gels with calcium salts or acids. It functions as a thickener, binder, and gelling agent in dessert gels, puddings, sauces, toppings, and edible films.

**Sodium Aluminosilicate—*See Sodium Silicoaluminate*.**

**Sodium Aluminum Phosphate, Acidic**—A leavening agent, slowly soluble in water, which gives it a delayed leavening reaction. It has a pH of 2.8. Approximately 20 percent of the carbon dioxide is released during the mixing period and the remainder is released during the baking period when the batter is exposed to heat. It has a high tolerance to variation in batter preparation. It is used in prepared mixes such as cake and pancake mixes.

**Sodium Aluminum Phosphate, Basic**—An emulsifier that is a white powder which is barely soluble in water. It has a pH of 9.2. It may be used in processed cheese to provide consistency and to aid in eliminating surface crystals.

**Sodium Aluminum Sulfate**—A leavening agent that releases the majority of the gas during baking, and is not used alone but in combination with a faster-acting leavening agent such as monocalcium phosphate. This results in a double-acting baking powder. It is almost nonreactive until heat is applied. It is used in baked goods.

**Sodium Ascorbate**—An antioxidant that is the sodium form of ascorbic acid. It is soluble in water and provides a non-acidic taste. A 10 percent aqueous solution has a pH of 7.3 to 7.6. In water, it readily reacts with atmospheric oxygen and other oxidizing agents, making it valuable as an antioxidant. One part sodium ascorbate is equivalent to 1.09 parts of sodium erythorbate. *See Ascorbic Acid*.

**Sodium Benzoate**—A preservative that is the sodium salt of benzoic acid. It converts to benzoic acid, which is the active form. It has a solubility in water of 50 g in 100 ml at 25°C. Sodium benzoate is 180 times as soluble in water at 25°C as is the parent acid. The optimum functionality occurs between pH 2.5 to 4.0 and it is not recommended above pH 4.5. It is active against yeasts and bacteria. It is used in acidic foods such as fruit juices, jams, relishes, and beverages. Its use level ranges from 0.03 to 0.10 percent.

**Sodium Benzosulfimide—*See Saccharin*.**

**Sodium Bicarbonate**—A leavening agent with a pH of approximately 8.5 in a 1 percent solution at 25°C. It functions with food grade phosphates (acidic leavening compounds) to release carbon dioxide which expands during the baking process to provide the baked good with increased volume and tender eating qualities. It is also used in dry-mix beverages to obtain carbonation, which results when water is added to the mix containing the sodium bicarbonate and an acid. It is a component of baking powder. It is also termed baking soda, bicarbonate of soda, sodium acid carbonate, and sodium hydrogen carbonate.

**Sodium Biphosphate**—*See Monosodium Phosphate.*

**Sodium Bisulfite**—A preservative that exists as a powder, with a solubility of 1 g in 4 ml of water. It prevents discoloration and inhibits bacterial growth. It is used in dried fruit to inhibit browning and maintain the bright color. It is found in reconstituted lemon juice. *See Sulfur Dioxide.*

**Sodium Calcium Aluminosilicate, Hydrated**—An anticaking agent for use at levels not to exceed 2 percent. It is also termed sodium calcium silicoaluminate.

**Sodium Calcium Silicoaluminate**—*See Sodium Calcium Aluminosilicate, Hydrated.*

**Sodium Caprate**—The sodium salt of capric acid. It functions as a binder, emulsifier, and anticaking agent.

**Sodium Caprylate**—The sodium salt of caprylic acid. It functions as a binder, emulsifier, and anticaking agent.

**Sodium Carbonate**—An alkali that exists as crystals or crystalline powder and is readily soluble in water. It has numerous functions: an antioxidant, a curing and pickling agent, a flavoring agent, a processing aid, a sequestrant, and an agent for pH control. It is used in instant soups to neutralize acidity. It is used in alginate water dessert gels to sequester the calcium, allowing the alginate to solubilize. It is also used in puddings, sauces, and baked goods.

**Sodium Carboxymethylcellulose**—*See Carboxymethylcellulose.*

**Sodium Carrageenan**—*See Carrageenan.*

**Sodium Caseinate**—The sodium salt of casein, a milk protein. It is used as a protein source and for its functional properties such as water binding, emulsification, whitening, and whipping. It is used in coffee whiteners, nondairy whipped toppings, processed meat, and desserts.

**Sodium Chloride**—*See Salt.*

**Sodium Citrate**—A buffer and sequestrant obtained from citric acid as sodium citrate anhydrous and as sodium citrate dihydrate or sodium citrate hydrous. The crystalline products are prepared by direct crystallization from aqueous solutions. Sodium citrate anhydrous has a solubility in water of 57 g in 100 ml at 25°C, while sodium citrate dihydrate has a solubility of 65 g in 100 ml at 25°C. It is used as a buffer in carbonated beverages and to control pH in preserves. It improves the whipping properties in cream and prevents feathering of cream and nondairy coffee whiteners. It provides emulsification and solubilizes protein in processed cheese. It prevents precipitation of solids during storage in evaporated milk. In dry soups, it improves rehydration which reduces the cooking time. It functions as a sequestrant in puddings. It functions as a complexing agent for iron, calcium, magnesium, and aluminum. Typical usage levels range from 0.10 to 0.25 percent. It is also termed trisodium citrate.

**Sodium Diacetate**—A preservative, sequestrant, acidulant, and flavoring agent that is a molecular compound of sodium acetate and acetic acid which yields acetic acid. It is a white crystalline powder which is hygroscopic. It functions against mold and bacteria and is used in bread. It is also termed sodium hydrogen diacetate.

**Sodium Dioctylsulfosuccinate**—*See Dioctyl Sodium Sulfosuccinate.*

**Sodium Erythorbate**—An antioxidant that is the sodium salt of erythorbic acid. In the dry crystal state it is nonreactive, but in water solution it readily reacts with atmospheric oxygen and other oxidizing agents, a property that makes it valuable as an antioxidant. During preparation, a minimal amount of air should be incorporated and it should be stored at a cool temperature. It has a solubility of 15 g in 100 ml of water at 25°C. On a comparative basis, 1.09 parts of sodium erythorbate are equivalent to 1 part of sodium ascorbate; 1.23 parts of sodium erythorbate are equivalent to 1 part erythorbic acid. It functions to control oxidative color and flavor deterioration in a variety of foods. In meat curing, it controls and accelerates the nitrite curing reaction and maintains the color brightness. It is used in frankfurters, bologna, and cured meats and is occasionally used in beverages, baked goods, and potato salad. It is also termed sodium isoascorbate.

**Sodium Ferric Pyrophosphate**—*See Sodium Iron Pyrophosphate.*

**Sodium Ferrocyanide**—*See Yellow Prussiate of Soda.*

**Sodium Hexametaphosphate**—A sequestrant and moisture binder that is very soluble in water but dissolves slowly. Solutions have a pH of 7.0. It permits peanuts to be salted in the shell by making it possible for the salt brine to penetrate the peanuts. In canned peas and lima beans, it functions as a tenderizer when added to the water used to soak or scald the vegetables prior to canning. It improves whipping properties in whipping proteins. It functions as a sequestrant for calcium and magnesium, having the best sequestering power of all the phosphates. It prevents gel formation in sterilized milk. It is also termed sodium metaphosphate and Graham's salt.

**Sodium Hydrogen Carbonate**—*See Sodium Bicarbonate.*

**Sodium Hydrogen Diacetate**—*See Sodium Diacetate.*

**Sodium Hydrogen Malate**—An acidulant.

**Sodium Hydroxide**—An alkali that is soluble in water, having a solubility of 1 g in 1 ml of water. It is used to destroy the bitter chemicals in olives that are to become black olives. It also functions to neutralize acids in various food products.

**Sodium Hypophosphite**—An emulsifier or stabilizer that is a white, odorless, deliquescent granular powder with a saline taste. It is also prepared as colorless, pearly crystalline plates. It is soluble in water, alcohol, and glycerol. It is prepared by neutralization of hypophosphorous acid or by direct aqueous alkaline hydrolysis of white phosphorus.

**Sodium Hyposulfite**—*See Sodium Thiosulfate.*

**Sodium Iron EDTA**—*See Iron.*

**Sodium Iron Pyrophosphate**—A nutrient and dietary supplement that is a source of iron. It contains approximately 14.5 percent iron and is insoluble in water. It is utilized for the enrichment of foods that are susceptible to rancidity. It is also termed sodium ferric pyrophosphate.

**Sodium Isoascorbate**—*See Sodium Erythorbate.*

**Sodium Lactate**—A humectant that is the sodium salt of lactic acid which is low melting and hygroscopic with a mildly saline taste. It is used in sponge cake and Swiss roll to produce a tender crumb and to reduce staling. It provides a protein plasticizing effect in biscuits. It is used in frankfurter-type sausages as a replacement for sodium chloride to extend shelf life and as a dehydrating salt or humectant in uncured hams. It can function as a flavoring agent and enhancer in some meat and poultry products.

**Sodium Laurate**—The sodium salt of lauric acid. It functions as a binder, emulsifier, and anticaking agent.

**Sodium Lauryl Sulfate**—An emulsifier and whipping aid that has a solubility of 1 g in 10 ml of water. It functions as an emulsifier in egg whites. It is used as a whipping aid in marshmallows and angel food cake mix. It also functions to aid in dissolving fumaric acid.

**Sodium Metabisulfite**—A preservative and antioxidant that exists as crystals or powder having a sulfur dioxide odor. It is readily soluble in water. It is used in dried fruits to preserve flavor, color, and to inhibit undesirable microorganism growth. It prevents "black spots" due to oxidative deterioration in shrimp. It is used in maraschino cherries. It is found in lemon drinks as a preservative. *See Sulfur Dioxide*.

**Sodium Metaphosphate**—*See Sodium Hexametaphosphate*.

**Sodium Myristate**—The sodium salt of myristic acid. It functions as a binder, emulsifier, and anticaking agent.

**Sodium Nitrate**—The salt of nitric acid that functions as an antimicrobial agent and preservative. It is a naturally occurring substance in spinach, beets, broccoli, and other vegetables. It consists of colorless, odorless crystals or crystalline granules. It is moderately deliquescent in moist air and is readily soluble in water. It is used in meat curing to develop and stabilize the pink color. *See Nitrate*.

**Sodium Nitrite**—The salt of nitrous acid that functions as an antimicrobial agent and preservative. It is a slightly yellow granular powder or nearly white, opaque mass or sticks. It is deliquescent in air. It has a solubility of 1 g in 1.5 ml of water. It is used in meat curing for color fixation and development of flavor. *See Nitrite*.

**Sodium Oleate**—The sodium salt of oleic acid. It functions as a binder, emulsifier, and anticaking agent.

**Sodium Palmitate**—The sodium salt of palmitic acid. It functions as a binder, emulsifier, and anticaking agent.

**Sodium Phosphate, Dibasic**—*See Disodium Phosphate*.

**Sodium Phosphate, Dibasic Dihydrate**—*See Disodium Phosphate*.

**Sodium Phosphate, Monobasic—***See Monosodium Phosphate*.

**Sodium Phosphate, Tribasic—***See Trisodium Phosphate*.

**Sodium Potassium Tartrate**—A buffer and sequestrant that is the salt of L(+)—tartaric acid. It has a solubility in water of 1 g in 1 ml. It is also termed Rochelle salt and potassium sodium tartrate.

**Sodium Propionate**—An antimicrobial agent that is the sodium salt of propionic acid. It occurs as colorless, transparent crystals or a granular crystalline powder. It is odorless or has a faint acetic-butyric acid odor, and is deliquescent. It is prepared by neutralizing propionic acid with sodium hydroxide. It is used in baked goods; nonalcoholic beverages; cheeses; confections and frostings; gelatins, puddings, and fillings; jams and jellies; meat products; and soft candy.

**Sodium Pyrophosphate—***See Sodium Acid Pyrophosphate*.

**Sodium Pyrophosphate, Tetrabasic—***See Tetrasodium Pyrophosphate*.

**Sodium Saccharin—***See Saccharin*.

**Sodium Sesquicarbonate**—A pH control agent that is prepared by: (1) partial carbonation of soda ash solution followed by crystallization, centrifugation, and drying; (2) double refining of trona ore, a naturally occurring impure sodium sequicarbonate. It is used in cream manufacture at a level of the ingredient sufficient to control lactic acid prior to pasteurization and churning of cream into butter.

**Sodium Silicate**—A product used as a preservative for eggs.

**Sodium Silicoaluminate**—An anticaking and conditioning agent used to improve flow properties and prevent caking. It absorbs moisture up to 75 percent of its weight. It functions as a moisture absorbent, moisture barrier, carrier, and processing aid. It is used in salt, cake mixes, powdered sugar, nondairy creamers, and dry mixes. Usage level ranges from 1 to 2 percent. It is also termed sodium aluminosilicate.

**Sodium Sorbate**—A preservative that is the salt of sorbic acid. It is partially soluble in water and is used effectively against yeasts and molds up to pH 6.5. It is not usually used as a replacement for sorbic acid or potassium sorbate. It is used in cheese and baked goods.

**Sodium Stearate**—The sodium salt of stearic acid. It functions as a binder, emulsifier, and anticaking agent. It is used as a plasticizer in chewing gum base.

**Sodium Stearyl Fumarate**—A dough conditioner and conditioning agent that is a white powder practically insoluble in water. It is used as a dough conditioner in yeast-raised baked goods. It is used as a conditioning agent in dehydrated potatoes. It also functions as a maturing and bleaching agent.

**Sodium Stearyl Lactylate**—A dough conditioner, emulsifier, and whipping agent that is the reaction product of stearic and lactic acids neutralized to a sodium or calcium salt, for example, calcium stearyl lactylate and sodium stearyl lactylate. It is used to improve the tolerance of bread dough to processing and to improve gas retention. It is used as an emulsifier in coffee whiteners, puddings, and low-fat margarine. It functions as a whipping aid in egg products and vegetable fat toppings. It complexes starch in dehydrated potatoes to allow for production of thicker, more uniform sheets.

**Sodium Sulfate**—The salt of sulfuric acid that is readily soluble in water and exists as crystals or crystalline powder. It is used in caramel production.

**Sodium Sulfite**—*See Sulfur Dioxide.*

**Sodium Tartrate**—A sequestrant and stabilizer that is the disodium salt of L(+)—tartaric acid. It is soluble in water. It functions as a sequestrant and stabilizer in meat products and sausage casings. It is also termed disodium tartrate.

**Sodium Tetrametaphosphate**—A sequestrant and emulsifier that is infinitely soluble in water. It is used as a water binder in cured pork. It is also termed Graham's salt.

**Sodium Thiosulfate**—A sequestrant, antioxidant, and formulation aid that is a powder soluble in water. It can be used in alcoholic beverages at 5 parts per million and in table salt at 0.1 percent. It is also termed sodium hyposulfite.

**Sodium Triphosphate**—*See Sodium Tripolyphosphate.*

**Sodium Tripolyphosphate**—A binder, stabilizer, and sequestrant that is mildly alkaline, with a pH of 10, and moderately soluble in water, with a solubility of 15 g in 100 ml of water at 25°C. It is used to improve the whipping properties of egg-containing angel food cake mix and meringues. It reduces gelling of juices and canned ham and tenderizes canned peas and lima beans. It is a moisture binder in cured pork and protects against discoloration and reduces shrinkage in sausage products. In algin desserts, it functions as a calcium sequestrant. It is also termed pentasodium tripolyphosphate and sodium triphosphate.

**Softener**—A term used for ingredients that soften. Softening relates to the hygroscopicity and the ability of the polyhydric alcohol, such as propylene glycol or glycerin, to retain moisture. Softeners are used in shredded coconut, pet foods, and chewing gum to maintain moistness. It is also termed plasticizer.

**Sorbic Acid**—A preservative that is effective against yeasts and molds. It is effective over a broad pH range up to pH 6.5, being ineffective above pH 7.0. It is a white, free-flowing powder which is slightly soluble in water with a solubility of 0.16 g in 100 ml of water at 20°C. Its solubility in water increases with increasing temperatures, although it is not recommended in foods that are pasteurized because it breaks down at high temperatures. The salts are potassium, calcium, and sodium sorbate. It is used in cheese, jelly, beverages, syrup, and pickles. Typical usage levels range from 0.05 to 0.10 percent.

**Sorbitan Ester**—A lipophilic emulsifier whose permitted type in foods is sorbitan monostearate. It is used in cakes, chocolate, and coffee whitener. It is also termed sorbitan fatty acid ester.

**Sorbitan Fatty Acid Ester**—*See Sorbitan Ester*.

**Sorbitan Monostearate**—A lipophilic emulsifier that is a sorbitan fatty acid ester, being a sorbitol-derived analog of glycerol monostearate. It is a nonionic, oil-dispersible surface-active agent. It is used as a gloss enhancer in chocolate coatings; as a dispersant aid in coffee whiteners; to increase volume in cakes and icings; and often in combination with polysorbates. Typical usage level ranges from 0.30 to 0.70 percent.

**Sorbitol**—A humectant that is a polyol (polyhydric alcohol) produced by hydrogenation of glucose with good solubility in water and poor solubility in oil. It is approximately 60 percent as sweet as sugar, and has a caloric value of 2.6 Kcal/gram. It is highly hygroscopic and has a pleasant, sweet taste. It maintains moistness in shredded coconut, pet foods, and candy. In sugarless frozen desserts, it depresses the freezing point, adds solids, and contributes some sweetness. It is used in low-calorie beverages to provide body and taste. It is used in dietary foods such as sugarless candy, chewing gum, and ice cream. It is also used as a crystallization modifier in soft sugar-based confections.

**Sorghum Oil**—An oil consisting mainly of linoleic and oleic fatty acids. It is similar in composition and properties to corn oil.

**Soybean**—A legume of high protein content, containing 40 percent or greater protein and approximately 18 percent oil. The protein contains all the essential amino acids. Soybeans are processed to produce soybean flour, protein concentrate, protein isolate, and soybean oil.

**Soybean Flour**—The flour made from defatted soybean, having a protein content in excess of 50 percent. It is used in doughnuts, cereal, bread, and sausage products for protein fortification and binding.

**Soybean Oil**—The oil obtained from the seed of the soybean legume. It consists of approximately 86 percent unsaturated fatty acids with linoleic and oleic being the principal two fatty acids. It exists in hydrogenated and unhydrogenated forms. It is used in shortenings and margarine in the hydrogenated form. It has some use in salad and cooking oils in the unhydrogenated form, but is limited by its tendency to develop undesirable odor and flavor when in contact with air or when heated to frying temperatures. It is also termed soy oil.

**Soybean Protein**—The protein obtained from soybeans, containing the essential amino acids. The most common forms are soybean flour (approximately 50 percent protein), soybean concentrate (approximately 70 percent protein), and soybean protein isolate (approximately 90 percent protein). It is used in sausages, snack foods, and meat analogs to provide emulsification, binding, moisture control, texture control, and protein fortification. It is also termed soy protein.

**Soybean Protein Concentrate**—The concentrate obtained by processing soybean flour to remove the soluble carbohydrates. The protein content is approximately 70 percent. In the powder form, it is used in processed meat products and sausage products for moisture and fat binding as well as texture. In baby food, cereal, and snack food it provides protein fortification. In the granular form, it is used in ground meat food items for texture. It is also termed soy protein concentrate.

**Soybean Protein Isolate**—The isolate prepared from soybean flour by extracting the protein and precipitating it to yield a product of approximately 90 percent protein. It functions to increase the protein content in foods, to reduce shrinkage, and to provide structure and appearance by emulsifying, stabilizing, and binding the fat and water. It is used in frozen spaghetti and meatballs, whipped toppings, and snack foods. It is also termed isolated soy protein and soy protein isolate.

**Soy Flour**—The powdered product obtained from defatted soybean. It has 50 percent or more protein content. It is used in doughnuts, bread, cereals, and sausage products as a nutrient and binder.

**Soy Flour, Lecithinated**—*See Lecithinated Soy Flour.*

**Soy Flour, Textured**—*See Textured Soy Flour.*

**Soy Oil**—*See Soybean Oil.*

**Soy Protein**—*See Soybean Protein.*

**Soy Protein Concentrate**—*See Soybean Protein Concentrate.*

**Soy Protein Isolate**—*See Soybean Protein Isolate.*

**Spice**—A variety of dried plant products that exhibit an aroma and flavor and from which no volatiles or other flavoring principles have been removed.

**Spirit Vinegar**—The product made by the acetous fermentation of dilute distilled alcohol, containing not less than 4 g of acetic acid per 100 cm$^3$ at 20°C. It functions as an acidulant and provider of flavor. It is used in mayonnaise, sauces, and salad dressings. It is also termed distilled vinegar and grain vinegar.

**Stabilizers and Thickeners**—Substances used to produce viscous solutions or dispersions, to impart body, improve consistency, or stabilize emulsions, including suspending and bodying agents, setting agents, jellying agents, and bulking agents, etc.

**Stannous Chloride**—An antioxidant and preservative that exists as white or colorless crystals, being very soluble in water. It reacts readily with oxygen, preventing its combination with chemicals and foods which would otherwise result in discoloration and undesirable odors. It is used for color retention in asparagus at less than 20 parts per million. It is also used in carbonated drinks.

**Starch**—A carbohydrate consisting of glucose units containing amylose and amylopectin which contribute to varying starch properties. Starch is insoluble in cold water, but upon heating the starch granules swell and burst forming starch paste. Starch sources include arrowroot, corn, potato, rice, sage, tapioca, waxy corn, and wheat. Starches are modified by treatment to alter their functional properties. Terminology designating these starches includes acid-modified cornstarch, food starch modified, modified food starch, oxidized cornstarch, pregelatinized starch, thin-boiling starch, and wheat starch, gelatinized. See specific starch.

**Stearic Acid**—A fatty acid that is a mixture of solid organic acids obtained principally from stearic acid and palmitic acid. It is practically insoluble in water. It functions as a lubricant, binder, and defoamer. It is used as a softener in chewing gum base.

**Stearyl Citrate**—An antioxidant made by reacting citric acid, which is not soluble in fats and oils, with stearyl alcohol, which readily dissolves in oils, thus enabling the citrate to dissolve in oil. It prevents metal ions from catalyzing oxidative reactions which cause rancidity. It is related to isopropyl citrate. It is used in vegetable oils and margarines.

**Stearyl Lactylate**—A dough conditioner, emulsifier, and whipping agent that is the reaction product of stearic and lactic acid neutralized to a sodium or calcium salt, for example, calcium stearyl lactylate and sodium stearyl lactylate. It is used to improve the tolerance of bread dough to processing and to improve gas retention. It is used as an emulsifier in coffee whiteners, puddings, and low-fat margarine. It functions as a whipping aid in egg products and vegetable fat toppings. It complexes starch in dehydrated potatoes to allow for production of thicker, more uniform sheets.

**Stearyl Monoglyceridyl Citrate**—An emulsion stabilizer prepared by controlled chemical reaction of citric acid, monoglycerides of fatty acids, and stearyl alcohol. It is used in or with shortenings containing emulsifiers.

**Stearyl Propylene Glycol Hydrogen Succinate**—*See Succistearin*.

**Sterculia Gum**—*See Karaya*.

**Straight Flour**—All of the flour that can be milled from a wheat blend, or approximately 72 percent of the wheat kernel which equates to 100 percent separation.

**Succinic Acid**—An acidulant that is commercially prepared by the hydrogenation of maleic or fumaric acid. It is a nonhygroscopic acid but is more soluble in 25°C water than fumaric and adipic acid. It has low acid strength and slow taste build-up; it is not a substitute for normal acidulants. It combines with proteins in modifying the plasticity of bread dough. It functions as an acidulant and flavor enhancer in relishes, beverages, and hot sausages.

**Succinic Anhydride**—An acidulant that hydrolyzes very slowly to succinic acid in water. It has thermal stability and a low melting point of 118°C which permits it to be used in products at comparatively low temperatures. It is used as a leavening acidulant for baking powder.

**Succinylated Monoglycerides**—Emulsifiers and dough conditioners made by the dissociation of succinylated monoglycerides. They are used in baked goods at 0.056 to 0.113 kg per 45.4 kg of flour to provide dough strength, improve shelf life, and improve texture. They are also used in shortenings.

**Succistearin**—An emulsifier that is the reaction product of succinic anhydride, fully hydrogenated vegetable oil (predominantly C16 or C18 fatty acid chain length), and propylene glycol. It is used in or with shortenings and edible oils intended for use in cakes, cake mixes, fillings, icings, pastries, and toppings. It is also termed stearyl propylene glycol hydrogen succinate.

**Sucralose**—High intensity sweetener manufactured by replacing three hydroxyl groups on the sucrose molecule with three chlorine atoms. The results are a sweetener of 0 calories that is not digested. It is 600 times as sweet as sugar with a similar flavor profile. It is heat stable, readily soluble, and maintains its stability at elevated temperatures. It has been approved for use in specific categories that include baked products, beverages, confectioneries, and certain desserts and toppings.

**Sucrose**—*See Sugar*.

**Sucrose Acetate Isobutyrate**—Weighting agent for beverages made from sugar. It increases the specific gravity of flavoring oils used in citrus beverages to prevent separation. It is odorless and flavorless at usage levels. It is also termed SAIB.

**Sucrose Fatty Acid Esters**—Emulsifiers, texturizers that are the mono-, di-, and triesters of sucrose with fatty acids and are derived from sucrose and edible tallow, or hydrogenated edible tallow or edible vegetable oils. Ethyl acetate or methyl ethyl ketone or dimethyl sulfoxide and isobutyl alcohol (2-methyl-1-propanol) may be used in the preparation of sucrose fatty acid esters. Sucrose fatty acid esters may be used as follows: as emulsifiers in baked goods and baking mixes, in dairy product analogs, in frozen dairy desserts and mixes, and in whipped milk products; as texturizers in biscuit mixes; as components of protective coatings applied to fresh apples, avocados, bananas, banana plantains, limes, melons (honeydew and cantaloupe), papaya, peaches, pears, pineapples, and plums to retard ripening and spoiling.

**Sugar**—A sweetener that is the disaccharide sucrose, consisting of one molecule of glucose and one molecule of fructose. It is obtained as cane or beet sugar. It has relatively constant solubility and is a universal sweetener because of its intense sweetness and solubility. It is available in various forms which include granulated, brown, and powdered. It is used in desserts, beverages, cakes, ice cream, icings, cereals, and baked goods. It is also termed beet sugar, cane sugar, sacharose, and sucrose.

**Sugar Beet Extract Flavor Base**—A flavor that is the concentrated residue of soluble sugar beet extractives from which sugar and glutamic acid have been recovered, and which has been subjected to ion exchange to minimize the concentration of naturally occurring trace minerals. It is used as a flavor in food.

**Sugar, Brown**—*See Brown Sugar*.

**Sugar, Fruit**—*See Fructose*.

**Sugar, Natural**—*See Turbinado Sugar*.

**Sugar, Powdered**—*See Powdered Sugar*.

**Sugar, Raw**—A natural sugar that has been washed to remove the impurities. It has a light golden color resulting from the molasses and a larger crystal size than granulated sugar. It is used where the flavor of natural sugar is desired, such as in cookies, bread, and cakes.

**Sugar, Reducing**—*See Reducing Sugar*.

**Sugar, Superfine**—*See Superfine Sugar.*

**Sugar Syrup**—A sweetener that is clear solutions of sucrose existing in varying grades. There is a water-white grade which is a sparkling clear syrup used in canned goods and beverages. There is also a light straw grade which has small amounts of color and nonsugar components.

**Sugar Syrup, Invert**—*See Invert Sugar Syrup.*

**Sugar, Washed Raw**—*See Turbinado Sugar.*

**Sulfur Dioxide**—A preservative, being a gas that dissolves in water to yield sulfurous acid. Sulfite salts, such as sodium and potassium sulfite, sodium and potassium bisulfite, and sodium and potassium metabisulfite, yield free sulfurous acid at low pH. Sulfur dioxide prevents the discoloration of foods by combining with the sugars and enzymes. It also inhibits bacterial growth. It is used in beverages, cherries, wines, and fruits.

**Sulfuric Acid**—An acidulant that is a clear, colorless, odorless liquid with great affinity for water. It is prepared by reacting sulfur dioxide with oxygen and mixing the resulting sulfur trioxide with water, or by reacting nitric oxide with sulfur dioxide in water. It is very corrosive. It is used as a modifier of food starch and is used in caramel production and in alcoholic beverages.

**Sunflower Oil**—A highly polyunsaturated oil obtained from sunflower seeds. There are two types of sunflower grown: an oilseed type used as a vegetable oil, and a non-oilseed type used for human food and bird seed. The composition of sunflower oil varies according to location and growing temperature. In general, sunflowers grown above the 39th parallel are high in linoleic acid and those grown below are high in oleic acid. The high linoleic variety is used for margarine and salad oil, while the high oleic variety is used in frying applications. This bland-flavored oil has a smoke point of 485 to 490°F (252 to 254°C) which gives it utility in baking, cooking, and frying foods. It is also used as a salad oil. In the hydrogenated form, it is used in margarine and shortenings.

**Sunset Yellow FCF**—*See FD&C Yellow #6.*

**Superfine Sugar**—Regular sugar ground into small crystals which increase the rate of dissolving. Used in beverages, finely textured cakes, and fruits.

**Surface-Active Agents**—Agents used to modify surface properties of liquid food components for a variety of effects, other than emulsification. Agents include solubilizing agents, dispersants, detergents, wetting agents, rehydration enhancers, whipping agents, foaming agents, and defoaming agents.

**Surface-Finishing Agents**—Agents used to increase palatability, preserve gloss, and inhibit discoloration of foods, including glazes, polishes, waxes, and protective coatings. Examples include coumarone-indene resin, methyl esters of fatty acids produced from e fats and oils, microcapsules for flavoring substances, morpholine, oxidized polyethylene, petroleum naphtha, polyacrylamide, sulfated butyl oleate, synthetic paraffin and succinic derivatives, and terpene resin.

**Sweet Basil**—*See Basil*.

**Sweet Pepper**—*See Paprika*.

**Sweet Rice Flour**—*See Waxy Rice Flour*.

**Synthetic Glycerin Produced by Hydrogenolysis of Carbohydrates**—An emulsifier produced by the hydrogenolysis of carbohydrates may be safely used in food. It contains equal to or less than 0.2 percent by weight of a mixture of butanetriols.

**Synthetic Petroleum Wax**—A wax that is a mixture of solid hydrocarbons, parafinic in nature, prepared by catalytic polymerization of ethylene. Synthetic petroleum wax is used in chewing gum base as a masticatory substance, on cheese and raw fruits and vegetables as a protective coating, and as a defoamer in food.

# T

**Tallow**—Animal fat obtained by separation from connected tissue, usually in mutton or beef. It consists principally of oleic and palmitic acid. It is a source of fat and is used in cake mix. It is used mostly in shortening and cooking oils.

**Tangerine Oil, Expressed**—A flavoring agent that is a red, brown, or orange oily liquid with a pleasant aroma. Oil obtained from unripe fruit may be green. It is soluble in most fixed oils and mineral oil, slightly soluble in propylene glycol, insoluble in glycerin. It is obtained by expression of oils from peels of ripe fruit of Dancy tangerine and related varieties.

**Tannic Acid**—A sequestrant that refers to a mixture of hydrolyzable tannins of a more complex structure than gallic acid. It is used in clarifying beer and wine. *See Tannins*.

**Tannins**—These are phenolic compounds that have several hydrolyzable groups. They are classified as: (a) hydrolyzable, yielding phenols such as gallic acid in the presence of acid and heat; and (b) condensed, obtained from the extract of oak trees and not hydrolyzable. Tannins are used for taste and chemical properties and as a sequestrant. They affect the color and flavor of fruits and vegetables. They are used in fruits, wine, and beer to remove undesirable material by forming insoluble complexes with the proteins.

**Tapioca Starch**—Starch having a bland flavor, being opaque, and forming long, cohesive pastes. It is found mainly in the modified form, being the pearl, granulated form which has been treated to be less stringy. It is used in puddings and pie fillings.

**Taro**—The tropical tuber *Colocasia esculenta* which can be used to make poi. Poi is carbohydrate food made by cooking the underground stem (corm) of the taro plant. The corms must be cooked because the calcium oxalate crystals present in the raw vegetable will act as tiny needles in the mouth.

**Tarragon**—The dried leaves and flowering tops of the herb *Artemisia dracunculus* L. It has a distinct aroma and anise-like flavor. It is used in salads, fish, sauces, and vinegar.

142

**Tartaric Acid**—An acidulant that occurs naturally in grapes. It is hygroscopic and rapidly soluble, with a solubility of 150 g in 100 ml of distilled water at 25°C. It has a slightly tarter taste than citric acid, with a tartness equivalent of 0.8 to 0.9. It augments the flavor of fruits in which it is a natural constituent. It is used in grape- and lime-flavored beverages and grape-flavored jellies. It is used as an acidulant in baking powder and as a synergist with antioxidants to prevent rancidity.

**Tartrazine**—*See FD&C Yellow #5.*

**Terpene Resin**—A moisture barrier that is the betapinene polymer obtained by polymerizing terpene hydrocarbons derived from wood. It is used on soft gelatin capsules in an amount not to exceed 0.07 percent of the weight of the capsule, and on powders of ascorbic acid or its salts in an amount not to exceed 7 percent of the weight of the powder.

**Tertiary Butylhydroquinone**—(TBHQ) An antioxidant that exhibits an excellent stabilizing effect in unsaturated fats and oils. It has good solubility in fats and oils, with a maximum usage level of 0.02 percent based on the weight of the fat or oil or the fat content of the food product. It shows no discoloration in the presence of iron and produces no discernible flavor or odor. It can be combined with BHA and BHT. It is used in edible fats and vegetable oils to retard rancidity. It is used in potato chips and dry cereal. It is also termed butylhydroquinone and mono-tertiary-butylhydroquinone.

**Tetrahydrofurfuryl Acetate**—A synthetic flavoring agent that is a stable, colorless liquid of slightly fruity odor. It should be stored in glass or tin containers. It is used in fruit flavors for application in candy and baked goods at 2 to 20 parts per million.

**Tetrahydrofurfuryl Propionate**—A synthetic flavoring agent that is a stable, colorless liquid of chocolate note. It should be stored in glass or tin-lined containers. It is used in flavors for chocolate with applications in beverages and ice cream at 2 parts per million, and in candy and baked goods at 20 parts per million.

**Tetrasodium Diphosphate**—*See Tetrasodium Pyrophosphate.*

**Tetrasodium Pyrophosphate**—A coagulant, emulsifier, and sequestrant that is mildly alkaline, with a pH of 10. It is moderately soluble in water, with a solubility of 0.8 g per 100 ml at 25°C. It is used as a coagulant in noncooked instant puddings to provide thickening. It functions in cheese to reduce the meltability and fat separation. It is used as a dispersant in malted milk and chocolate drink powders. It prevents crystal formation in tuna. It is also termed sodium pyrophosphate, tetrasodium diphosphate, and TSPP.

**Textured Soy Flour**—Soy flour that is processed and extruded to form products of specific texture and form, such as meatlike nuggets. The formed products are crunchy in the dry form and upon hydration become moist and chewy.

**Textured Vegetable Protein**—A vegetable protein that is processed and extruded to form beeflike strips, meatlike nuggets, or other analogs. In the dehydrated form, the analogs are crunchy and upon hydration become moist and chewy. Soy protein is the most popular protein source, but other vegetable proteins include peanut and wheat. It is used as meat analogs. It is also termed textured soy flour or textured soy protein.

**Thaumatin**—A flavor enhancer that is a protein which is approximately 3,000 times as sweet as sucrose. The onset of sweetness may take several seconds and can be affected by heat. It is used in chewing gum.

**THBP**—An antioxidant (2,4,5-trihydroxybutyrophenone) that is used alone or in combination with other permitted antioxidants. The total antioxidant content of a food containing the additive will not exceed 0.02 percent of the oil or fat content of the food, including the essential (volatile) oil content of the food.

**Thiamine**—The water-soluble vitamin $B_1$, required for normal digestion and functioning of nerve tissues and in the prevention of beriberi. It also acts as a coenzyme in the metabolism of carbohydrates. During processing, the higher and longer the heating period, the greater the loss. The loss is reduced in the presence of acid. Thiamine hydrochloride and thiamine mononitrate are two available forms. The mononitrate form is less hygroscopic and more stable than the hydrochloride form, making it suitable for use in beverage powders. It is used in enriched flour and is found as thiamine mononitrite in frozen egg substitute and crackers.

**Thiamine Mononitrate**—*See Thiamine.*

**Thin-Boiling Starch**—*See Cornstarch, Acid-Modified.*

**Thiodipropionic Acid**—An antioxidant used to prevent fats and oils from going rancid. It has the same functionality as BHA, BHT, and propyl gallate.

**Thyme**—The dried leaves and flowering tops of the shrub *Thymus vulgaris* L. There are two important variations: French thyme, which has a narrow leaf; and lemon thyme, which has a variegated leaf. It is used in soups, cheese, sauces, and appetizers.

**Titanium Dioxide**—A white pigment that disperses in liquids and possesses great opacifying power. The crystalline modifications of titanium dioxide are rutile and anatase, of which only anatase finds use as a color additive.

**Tocopherol**—Fat-soluble vitamin E, which is a light yellow oil readily degradable by heat. As a vitamin, it is essential for normal muscle growth and prevents vitamin A destruction by deterioration. It also functions as an antioxidant. It prevents the oxidation of certain fatty acids and is stable unless the food becomes rancid. Vegetable oils contain a higher concentration of natural antioxidants, including tocopherols, than animal fats and are thus more stable. Tocopherol is obtained from vegetable oils, beans, eggs, and milk. It is also termed alpha-tocopherol.

**Tofu**—A soybean curd product. Soybeans are soaked, ground, and filtered, with the remaining mixture being heated to 75°C and a coagulant added, which results in the formation of the soy curd and whey. The soy curd is pressed to separate it from the whey and then washed and cooled. It is low in calories and saturated fats while high in vitamins, minerals, and digestible protein. It is tasteless, but takes on the flavors of the products with which it is cooked. Uses include frozen desserts and meat products.

**Tomato Paste**—The paste prepared from tomatoes which are processed by heat to prevent spoilage. The paste contains not less than 24 percent tomato soluble solids.

**Tragacanth**—A gum produced from a bush of the genus *Astragalus*. It swells in water to give a highly viscous sol or paste. A 1 percent solution of the purest gum has a viscosity of approximately 3400 cps, and about 2 percent can form a paste. Solutions have a pH of 5 to 6. It is stable at low pH and is an effective suspending agent because of its stability and acid resistance. It is used in salad dressings, sauces, fruit fillings, and citrus beverages. It is also termed gum tragacanth.

**Triacetin**—*See Glyceryl Triacetate.*

**Tribasic Calcium Phosphate**—*See Tricalcium Phosphate.*

**Tributyrin**—A flavoring agent that is the triester of glycerin and butyric acid. It is prepared by esterification of glycerin with excess butyric acid. It is used in the following foods: baked goods; alcoholic beverages; nonalcoholic beverages; fats and oils; frozen dairy desserts and mixes; gelatins, puddings and fillings; and soft candy. It is also termed butyrin and glyceryl tributyrate.

**Tricalcium Orthophosphate**—*See Tricalcium Phosphate.*

**Tricalcium Phosphate**—An anticaking agent and calcium source that is a white powder that is almost insoluble in water. It is used as an anticaking agent in table salt and dry vinegar. It is used as a source of calcium and phosphorus in cereals and desserts. It functions as a bleaching agent in flour and in lard, and prevents undesirable coloring and improves stability for frying. It is also termed tribasic calcium phosphate, tricalcium orthophosphate, calcium phosphate tribasic, and precipitated calcium phosphate.

**Tricalcium Silicate**—An anticaking agent used in table salt.

**Triethyl Citrate**—A sequestrant that is an oily liquid, slightly soluble in water. It is found in lemon drinks.

**Trihydroxybutyrophenone**—*See THBP.*

**Tripotassium Citrate**—*See Potassium Citrate, Monohydrate.*

**Trisodium Citrate**—A buffer and sequestrant that is the trisodium salt of citric acid. *See Sodium Citrate.*

**Trisodium Monophosphate**—*See Trisodium Phosphate.*

**Trisodium Orthophosphate**—*See Trisodium Phosphate.*

**Trisodium Phosphate**—An emulsifier and buffer that is strongly alkaline, with a pH of 12. It is moderately soluble in water, with a solubility of 14 g per 100 ml at 25°C. It functions as an emulsifier in processed cheese to improve texture. It maintains viscosity and prevents phase separation in evaporated milk and is also found in cereals. It is also termed trisodium orthophosphate, sodium phosphate tribasic, and trisodium monophosphate.

**Trisodium Phosphate Crystals**—An emulsifier and buffer with a solubility in water of 50 g per 100 ml at 25°C. It is used in processed cheese as an emulsifier and it is also used in denture cleaner formulations.

**Turbinado Sugar**—Washed raw sugar of light gold color and larger grain size than regular sugar. It has a thin film of molasses which contributes toward the distinctive flavor. It is also termed natural sugar and washed raw sugar.

**Turmeric**—A spice and colorant that is the rhizome or root of *Curcuma longa*. As a spice, it has a taste related to mustard. As a vegetable color, it has a bright yellow to greenish-yellow hue. The yellow pigment is curcumin. It is water miscible and has excellent heat stability, poor light and pH stability, and good tinctorial strength. It exists as an extract and oleoresin. It is used in meat, poultry, fish, and rice dishes.

# U

**(Gamma)-Undecalactone**—A synthetic flavoring agent that is a colorless to yellow liquid of strong peach fruit odor. It is unstable to alkali and stable to weak organic acids. It should be stored in glass or tin containers. It is used in flavors for its peach note and has application in gelatins, puddings, beverages, ice cream, and candy at 7 to 11 parts per million.

**Undecanal**—A flavoring agent that is a liquid, colorless or pale yellow, with a sweet odor. It is soluble in most fixed oils, mineral oil, and propylene glycol; insoluble in glycerin. It is obtained by chemical synthesis. It is also termed aldehyde C-11 undecyclic and *n*-undecyl aldehyde.

**Unmodified Cornstarch**—*See Cornstarch*.

# V

**Vanaspati**—A vegetable fat used in candy.

**Vanilla**—A flavorant obtained from the cured vanilla bean. The vanilla or vanilla bean refers to the fully grown, unripe, cured, and dried fruit pod of the vanilla vine *Vanilla planifolia*. Those beans produced in Madagascar and its neighboring islands are termed "Bourbon beans"; those produced in Indonesia are termed "Java beans." The bean contains 1.5 to 3.0 percent vanillin, the most powerful flavorant in the cured bean, along with approximately 10 percent of other extractives. It is used in the comminuted form in "Philadelphia" type ice cream or as a vanilla flavorant in sauces or liquids by suspending the whole bean in them. Most vanilla flavoring is done with vanilla extract.

**Vanilla Extract**—A flavorant made from vanilla bean extract. It is a solution containing not less than 35 percent alcohol of the components extracted from one or more units of vanilla constituent. One unit is 0.378 kg of vanilla beans containing not more than 25 percent moisture. A double-strength solution (2-fold) contains twice the quantity of beans. It is used in desserts, baked goods, and beverages.

**Vanilla Flavor, Artificial**—A flavorant that consists of vanilla reinforced with synthetic vanillin. The best imitation vanillas contain vanillin, ethyl vanillin, or very little coumarin with or without vanilla, while the poorer ones contain high levels of coumarin. It is used in desserts, baked goods, and beverages.

**Vanilla Sugar**—A flavorant consisting of sugar mixed with vanilla extract. It is used in desserts and other food products.

**Vanillin**—A flavorant made from synthetic or artificial vanilla which can be derived from lignin of whey sulfite liquors and is synthetically processed from guaiacol and eugenol. The related product, ethyl vanillin, has three and one-half times the flavoring power of vanillin. Vanillin also refers to the primary flavor ingredient in vanilla, which is obtained by extraction from the vanilla bean.

Vanillin is used as a substitute for vanilla extract, with application in ice cream, desserts, baked goods, and beverages at 60 to 220 parts per million.

**Vanillin Acetate**—A synthetic flavoring agent that is moderately stable, white to yellow crystals of vanilla odor. It should be stored in glass or polyethylene-lined containers. It is used in flavors for vanilla note, with application in beverages, ice cream, candy, and baked goods at 11 to 28 parts per million.

**Vegetable Gums**—Gums that are water thickeners obtained from a plant source.

**Vegetable Oil, Hydrogenated**—*See Hydrogenated Vegetable Oil*.

**Vegetable Oils**—Oils obtained from a vegetable source, including soy beans, peanuts, cottonseeds, and palms. They are used in cooking and salad oils and dressings.

**Vegetable Protein, Textured**—*See Textured Vegetable Protein*.

**Vinegar**—An acidulant and flavorant that, with regard to general types, is the product produced from cider, grapes, sucrose, glucose, or malt by successive alcoholic and acetous fermentations in which acetic acid is the principal measured component. The term *vinegar* applies only to cider vinegar, also termed apple vinegar. The acetic acid content is measured in grains, where 10 grains equals 1 percent acetic acid. It is used in salad dressings and sauces.

**Vinegar, Distilled**—The product made by the acetous fermentation of dilute distilled alcohol without addition of color, containing not less than 4 g of acetic acid per 100 cm$^3$ at 20°C. The acetic acid content is measured in grains, where 10 grains equal 1 percent acetic acid. It is used in mayonnaise and salad dressing. It is also termed spirit vinegar and grain vinegar.

**Vital Wheat Gluten**—A powder of high protein content obtained by drying freshly washed gluten under controlled temperature conditions. It absorbs approximately twice its weight of water and readily forms a cohesive, elastic dough. It is used in bread, rolls, and buns. *See Wheat Gluten*.

**Vitamin A**—*See Retinol*.

**Vitamin B$_1$**—*See Thiamine*.

**VitaminB₂—*See Riboflavin*.**

**Vitamin B₅—*See Pantothenic Acid*.**

**Vitamin B₆—*See Pyridoxine*.**

**Vitamin B₆ Hydrochloride—*See Pyridoxine Hydrochloride*.**

**Vitamin B₁₂—*See Cyanocobalamin*.**

**Vitamin C—*See Ascorbic Acid*.**

**Vitamin D₂—*See Calciferol*.**

**Vitamin E—*See Tocopherol*.**

**Vitamin K**—A fat-soluble vitamin that is essential for blood clotting. It is destroyed by irradiation during processing but has no appreciable loss during storage. It occurs in spinach, cabbage, liver, and wheat bran.

**Vitamins**—Organic compounds that are essential for normal body growth and maintenance. They are classified into two groups: fat-soluble (vitamins A, D, E, and K), and water-soluble (vitamins B and C). Vitamins are measured in very low concentrations, such as 1 to 100 mg. Through biochemical action, they perform various functions in such processes as cell growth, normal digestion, manufacture of red blood cells, and absorption of calcium and phosphorus. Inadequate vitamin intake can be the result of food deficiency, increased vitamin requirements, and increased vitamin loss. The vitamins of determined importance include: A (retinol), B₁ (thiamine), B₂ (riboflavin), B₅ (pantothenic acid), B₆ (pyridoxine), B₁₂ (cyanocobalamin), C (ascorbic acid), D₂ (calciferol), E (tocopherol), K, niacin, folic acid, and biotin.

# W

**Washed Raw Sugar**—*See Turbinado Sugar.*

**Water**—A colorless, odorless, tasteless liquid formed by the combination of two hydrogen and one oxygen atoms. It allows substances to dissolve and functions as a solvent, dispersing medium, hydrate, and promoter of chemical changes. It is a major constituent in meats, fruits, and vegetables. Distilled water is obtained by condensation of water vapor.

**Waxy Corn**—A corn consisting essentially of amylopectin (pure branched-chain polymers), which differentiates it from regular corn, which consists of amylose and amylopectin. The amylopectin content results in a starch which upon heating forms a clear, cohesive paste that does not form a true gel upon cooling. It has a high water-binding capacity and is resistant to gel formation and retrogradation. It is used in puddings and sauces.

**Waxy Maize Starch**—The starch portion of waxy corn, consisting essentially of amylopectin. It yields pastes that are almost clear upon cooling and are noncongealing. It forms a translucent, water-soluble coating when dried in thin films. It is used to thicken a variety of foods such as sauces and puddings. It is also termed waxy starch and amioca.

**Waxy Rice Flour**—A flour obtained from waxy rice, which contains almost no amylose. It is comparable in viscosity characteristics to waxy corn flour. It has less than 0.5 percent amylose in the starch and contains alpha-amylose. It has excellent resistance to syneresis during freeze/thaw cycles. It is used in frozen sauces and gravies. It is also termed sweet rice flour.

**Waxy Rice Starch**—A rice starch that contributes freeze-thaw stability to sauces and puddings but may provide objectionable flavor.

**Waxy Sorghum**—A type of sorghum characterized by having paste clarity, high water-binding capacity, and resistance to gel formation and retrogradation. The unmodified form results in a stringy, cohesive paste. It is used in dressings with other starches.

**Waxy Starch**—*See Waxy Maize Starch.*

**Wetting Agents**—*See Surface-Active Agents.*

**Wheat**—A cereal grain in which the kernel is separated by milling into flour, bran, and germ. It is used in all types of farinaceous foods. *See Flour; Wheat Flour.*

**Wheat, Bulgur**—*See Bulgur Wheat.*

**Wheat Flour**—A fine powdery substance obtained by milling wheat with application in farinaceous foods.

**Wheat Germ**—The oil-containing portion of the wheat kernel.

**Wheat Gluten**—The water-insoluble complex protein fraction separated from wheat flours. Gum gluten is wheat gluten in its freshly extracted wet form. Dry gluten is approximately 70 to 80 percent protein but is deficient in the amino acid lysine. It absorbs two to three times its weight in water. The differences in properties of wheat gluten in comparison to almost all other food proteins are largely due to the low polarity level of the total amino acid structure. Most food proteins have polar group levels of 30 to 45 percent and have a net negative charge, while wheat gluten has a polar group level of approximately 10 percent with a net positive charge. This results in the repulsion of excess water and the close association of the wheat gluten molecules and resistance to dispersion. In baked goods, this results in the ability to form adhesive, cohesive masses, films, and three-dimensional networks. Gluten formation is utilized in the baking industry to impart dough strength, gas retention, structure, water absorption, and retention with breads, cakes, doughnuts, and so on. It is also used as a formulation aid, binder, filler, and tableting aid. *See Gluten; Vital Wheat Gluten.*

**Wheat Starch**—A starch obtained from wheat. It produces lower viscosity and more tender gels than starch obtained from corn or sorghum. It has a gelatinization range about 10°C lower than corn or waxy maize starch. It is used in the baking industry to permit the use of hard wheat flour in baked goods. It functions as a binder in breading and batter mixes. It is used in soups, pie fillings, sauces, and gravies.

**Wheat Starch, Gelatinized**—*See Pregelatinized Starch.*

**Whey**—The portion of milk remaining after coagulation and removal of curd. There are two principal types: sweet whey obtained during the making of rennet-type hard cheeses like Cheddar and Swiss, with a pH of approximately 6.1; and acid whey obtained during the making of acid-type cheeses, such as cottage cheese, with a pH of approximately 4.4 to 4.6. Whey is used as a source of lactose, milk solids, and whey proteins. It is used in baked goods, ice cream, and dry mixes.

**Whey Protein Concentrate**—Concentrated whey which is obtained from cheese originally consisting of approximately 12 percent protein, 0.5 percent fat, and 65 to 70 percent lactose. The whey concentrate increases the protein content, usually ranging from 33 to 55 percent. Properties provided are water control, increase in viscosity, opacity, and network interruption as a fat replacer. Uses are as a fat replacement in cheese, frozen desserts, dairy products, and baked goods.

**Whey Solids**—The solid fraction or dry form of whey. It is used as a replacement for milk solids–not-fat to provide a source of protein, solids, and flavor. It is used in baked goods, ice cream, dry mixes, and beverages.

**Whole Fish Protein Concentrate**—A protein supplement that is derived from whole hake and hakelike fish, herring of the genera *Clupea*, menhaden, and anchovy of the species *Engraulis mordax*. The additive consists essentially of a dried fish protein processed from the whole fish without removal of heads, fins, tails, viscera, or intestinal contents. It is prepared by solvent extraction of fat and moisture with isopropyl alcohol or with ethylene dichloride followed by isopropyl alcohol, except that the additive derived from herring, menhaden, and anchovy is prepared by solvent extraction with isopropyl alcohol alone. Solvent residues are reduced by conventional heat drying and/or microwave radiation and there is a partial removal of bone.

**Whole Milk**—*See Milk.*

**Whole Milk Solids**—The product resulting from the drying or desiccation of milk. It contains not less than 26 percent fat and not more than 5 percent moisture. The dry form offers convenience of transportation, utility, and stability. It is used in dry mixes such as puddings, soup mixes, and desserts. It is also termed dried milk, dry whole milk, and milk powder.

**Whole Wheat Flour**—The flour obtained by grinding cleaned wheat, other than durum wheat or red durum wheat, with the proportions of the natural constituents, other than moisture, remaining unaltered. The moisture content is not more than 15 percent. Optional ingredients include malted wheat, wheat flour, and barley flour for compensation for any natural deficiency of enzymes; ascorbic acid; and bleaching ingredients. It is also termed graham flour and entire wheat flour.

**Wine Vinegar**—The vinegar made by the alcoholic and acetous fermentation of the juices of grapes or wine. It contains a minimum of 4 g per 100 cm$^3$ acid expressed as acetic acid. There is red wine vinegar, which has a rose to deep red color, and white wine vinegar, which has a pale yellow to off-white color. It is used in salad dressings, marinades, and sauces.

**Worcestershire Sauce**—A sauce consisting of water, vinegar, soy sauce, corn syrup, salt and spices, or variations of these ingredients. It is used as a flavorant and is found in barbeque sauce and sweet-and-sour sauces.

**X**

**Xanthan Gum**—A gum obtained by microbial fermentation from the *Xanthomonas campestris* organism. It is very stable to viscosity change over varying temperatures, pH, and salt concentrations. It is also very pseudoplastic which results in a decrease in viscosity with increasing shear. It reacts synergistically with guar gum to provide an increase in viscosity and with carob gum to provide an increase in viscosity or gel formation. It is used in salad dressings, sauces, desserts, baked goods, and beverages at 0.05 to 0.50 percent.

**Xylitol**—A polyhydric alcohol that is a natural sugar substitute commercially made from xylan-containing plants (birch) hydrolyzed to xylose. It is as sweet as sucrose, dissolves quickly, and has a negative heat of solution which results in a cooling effect. It has 24 Kcal/gram. It is used in chewing gum, throat lozenges, and chocolate.

# Y

**Yeast**—A leavening and fermentation agent that is a single-celled plant that can convert sugar to carbon dioxide. It is used as a leavening agent in bread and dough-type mixtures. It provides a yeasty flavor and tender crust. It has slow action as a leavening agent. One pound of active dry yeast replaces approximately 2 pounds of fresh yeast. Selected yeast strains are used in wine fermentation.

**Yeast Extract**—A flavor contributor and flavor enhancer consisting of a combination of nucleic acids, peptides, polypeptides, amino acids, and other constituents. It is obtained from the yeast cells of *Saccharomyces cerevisiae*, formed in the brewing of beer. It is used to provide the same functions as monosodium glutamate, although not to the same extent. It is used as a partial substitute for meat extract and also functions with other flavor ingredients such as hydrolyzed vegetable proteins. It is used in soups, gravies, spreads, dressings, and meat products. Typical usage levels range from 0.1 to 0.5 percent.

**Yeast Food**—A complete food used in doughs. It contains dough conditioner ingredients such as calcium salts, sulfates, and phosphates which strengthen the gluten. It also contains ammonium salts and phosphates which function as yeast nutrients. It is used in bread dough and in the fermentation of alcoholic beverages.

**Yeast-Malt Sprout Extract**—A food enhancer produced by partial hydrolysis of yeast extract (derived from *Saccharomyces cereviseae*, *Saccharomyces fragilis*, or *Candida utilis*) using the sprout portion of malt barley as the source of enzymes. The additive contains a maximum of 6 percent 5'nucleotides by weight.

**Yellow Prussiate of Soda**—An anticaking agent and crystallizing agent. It is sometimes added as a crystallizing agent to salt when it crystallizes to generate lagged and bulky crystals which resist caking. It also functions as a water-soluble anticaking agent. It is also termed sodium ferrocyanide.

**Yogurt**—A custard-like or soft gel product made by fermenting milk with bacterial cultures, specifically *Lactobacillus bulgaricus* and *Streptococcus thermophilus*, to a pH range of usually 4.0 to 4.5. It is used as a snack; as a meal; or in desserts, salad dressings, and baked goods.

**Yucca Plant Extract**—A foaming agent obtained from the yucca plant species *Yucca brevifolia* and *Yucca schidigera*. It is available in liquid concentrate or dried form, is dark brown, and has a slight bittersweet flavor and a pH of 4.0. It is stable over a wide pH range and heat treatment and is readily soluble in water. It is used in applications where a frothy appearance and foam stability are desired, such as in root beer, cocktail mixes, and whipped beverages. Usage level is 50 to 150 parts per million.

# Z

**Zein**—A corn protein produced from corn gluten meal. It lacks the amino acids lysine and tryptophan, so it is not suitable as a sole source of dietary protein. It is insoluble in water and alcohols but is soluble in aqueous alcohols, glycols, and glycol ethers. It functions as a film and coating to provide a moisture barrier for nuts and grain products. It also functions as a coating for confections and a glaze for panned goods.

**Zinc**—(Zn) A metallic element that functions as a nutrient and dietary supplement. It is believed to be necessary for nucleic acid metabolism, protein synthesis, and cell growth. Sources of zinc include zinc acetate, carbonate, chloride, citrate, gluconate, oxide, stearate, and sulfate. The gluconate form is used in lozenges. The sulfate form exists as prisms, needles, or powder. It has a solubility of 1 g in 0.6 ml of water and is found in frozen egg substitutes.

**Zinc Acetate**—*See Zinc.*

**Zinc Carbonate**—*See Zinc.*

**Zinc Chloride**—*See Zinc.*

**Zinc Citrate**—*See Zinc.*

**Zinc Gluconate**—*See Zinc.*

**Zinc Oxide**—*See Zinc.*

**Zinc Methionine Sulfate**—A source of dietary zinc that is the product of the reaction between equimolar amounts of zinc sulfate and DL-methionine in purified water. It is used in tablet form.

**Zinc Stearate**—*See Zinc.*

**Zinc Sulfate**—*See Zinc.*

# PART II

# Ingredient Categories

# ACIDULANTS

Acidulants are acids used in processed foods for a variety of functions that enhance the food. Acids are used as flavoring agents, preservatives in microbial control, chelating agents, buffers, gelling and coagulating agents, and in many other ways. Examples of these functions are:

- Flavoring agent—Contributes and enhances flavor in carbonated beverages, fruit drinks, and desserts.
- Preservative—An acid medium restricts the growth of spoilage organisms in mayonnaise and tomato sauce, and retards the activity of enzymes involved in discoloration in fruits.
- Chelating agent—Aids in binding metals that can cause oxidation in fats and oils, and discoloration in canned shrimp.
- Buffer—Maintains and controls acidity during processing, and maintains acidity within a given range in prepared desserts.
- Gelling agent—Controls the gelling mechanism of algin and pectin gels such as desserts and jams.
- Coagulating agent—Reduction of pH results in coagulation of milk protein which is used in preparation of direct acidified cheese and desserts.

Acidulant selection depends upon the application or processed food. The properties to consider are flavor profile, pH, solubility rate, solubility, and hygroscopicity. Flavor profile refers to the perceived sharpness or blandness contributed by the acid. Tartaric and citric acids provide a sharp taste as compared to lactic acid, which provides a blander taste. Fumaric acid and tartaric acid provide the greatest degree of sourness. Comparing acids relative to similarity of taste to citric acid, the relative equivalents are: citric acid: 100; fumaric acid: 55; tartaric acid: 70; malic acid: 75; succinic acid: 87; lactic acid: 107; and glucono-delta-lactone: 310. The acids provide different pHs at similar concentrations, falling generally in the range of pH 2 to 3 at 1 percent concentration. The solubility rate determines how rapidly the acid dissolves and contributes toward the flavor profile. In a beverage powder dissolved in cold water, rapid solubility is required, so perhaps citric acid would be used. The solubility of the acid refers to the quantity in solution. Cream of tartar and fumaric acid have low solubilities, which makes them suitable for bakery applications in controlling leavening systems. Phosphoric acid, a liquid, is miscible with water and used in beverage syrups. Hygroscopic acids will absorb moisture, and hygroscopicity needs to be considered when the application is dry mixes. If hygroscopic acids, such as citric or tartaric acids, are used in dry mixes, proper packaging is essential. Alternative less hygroscopic acids are adipic and fumaric acid.

Table 1 illustrates the properties of the principal acids used in the food industry.


## ANTIOXIDANTS

Antioxidants are chemical compounds that provide stability to fats and oils by delaying oxidation (which involves the loss of electrons, and the gain of oxygen). The oxidation of fats and oils is believed to occur as a series of chain reactions in which oxygen from the air is added to the free fat radical. The fat molecule loses a hydrogen atom and becomes an unstable free radical with a high affinity for oxygen. Oxygen is added and the fat molecule, to complete its electron structure, reacts with another fat molecule and removes a hydrogen atom. This produces another free radical and results in a chain reaction. The antioxidant functions by replacing the fat molecule as the hydrogen atom donor in order to complete the electron structure of the free radical, thus terminating the chain reaction. Thus oxidative rancidity, which results in off-flavors and odors, is retarded until the antioxidant supply is used.

The most commonly used antioxidant formulations contain combinations of BHA (butylated hydroxyanisole), BHT (butylated hydroxytoluene), and propyl gallate. These formulations usually contain a chelating agent (reacts with metal to form a complex and thus prevents the metal from acting as a catalyst in oxidative reactions), of which citric acid is the most common. Natural antioxidants such as the tocopherols and guaiac gum usually lack the potency of BHA, BHT, and propyl gallate combinations.

Antioxidants are effective at low concentrations, that is, 0.02 percent or less, based on the fat or oil content of the food. Examples of applications are:

- Rendered animal fat, such as lard
- Vegetable oils, such as cottonseed oil and corn oil
- Food products of high fat content, such as doughnuts and potato chips
- Food products of low fat content, such as cereals and dehydrated potatoes

**Table 1** Comparative Acid Chart

| Name | Chemical Formula | Solubility (g/100 ml Distilled H₂O at 20°C) | pH in H₂O — Percentage Solution | pH | Tartness | Comments |
|---|---|---|---|---|---|---|
| Adipic acid | $C_6H_{10}O_4$ | 1.9 | 0.6 1.2 | 2.86 | Smooth tart | Nonhygroscopic; imparts long-lasting flavor note |
| Citric acid, anhydrous | $C_6H_8O_7$ | 1.46 | 0.12 0.5 | 3.20 2.35 | Sharp tart | Immediate acid taste |
| Citric acid, monohydrate | $C_6H_8O_7 \cdot H_{20}$ | 175 | — | — | Sharp tart | Immediate acid taste |
| Fumaric acid | $C_4H_4O_4$ | 0.49 | 0.5 | 2.15 | Tart | Nonhygroscopic |
| Gluconic acid | $C_6H_{12}O_7$ | 100 | 1 | 2.80 | Mild | Unobtrusive |
| Glucono-delta-lactone | $C_6H_{10}O_6$ | 59 | See gluconic acid | | Mild | Low hygroscopicity; slowly converted to gluconic acid |
| Lactic acid | $C_3H_6O_3$ | Liquid | 5 | 2.25 | Smooth tart | Slow onset of acid taste which lingers |
| Malic acid | $C_4H_6O_5$ | 130 | 1 | 2.35 | Smooth tart | Hygroscopic; perception of sourness or sharpness, starts gradually, rises to peak and fades slowly |
| Phosphoric acid | $H_3PO_4$ | Liquid | 0.12 | 2.68 | — | Sour taste, high acidity |
| Succinic acid | $C_4H_6O_4$ | 5 | Low acid strength | | Tart | Nonhygroscopic; has slow flavor build-up |

Antioxidants should be added to fats and oils before oxidation has started in order to be effective. The antioxidant cannot reverse the oxidation process nor regenerate a product that has become rancid. The oxidation process is accelerated by heat, light, moisture, metals, and other factors.

Antioxidants include trihydroxybutyrophenone (THBP), dilauryl thiodipropionate (DLTDP), nordihydroguaiaretic acid (NDGA), guaiac gum, thiodipropionic acid, tocopherols, lecithin, sodium erythorbate, ascorbic acid, and ascorbyl palmitate.

## CHELATING AGENTS (SEQUESTRANTS)

Chelation is an equilibrium reaction between a chelating (complexing) agent and a metal ion which forms a complex. Trace metal ions in foods can produce undesirable effects such as discoloration, turbidity, and oxidation. The chelating agents can form a complex with the unwanted trace metals, thus blocking the reactive sites of the metal ions and rendering them inactive. The complex formed is termed a *chelate*, that is, metal + chelating agent = metal complex.

An equilibrium constant K defines the ratio of chelated metal to unchelated metal. The log K is the stability constant, measuring the affinity of the complexing agent for the metal ion. A high K value indicates a high affinity of the complexing agent for the metal ion and thus a low value for free metal ion concentration. Stability constants for some metal ions are shown in Table 2.

The most problematic metal ions in foods are iron and copper. In a system containing several metal ions, the ones with the highest

**Table 2** Metal Ion Stability Constants

| Metal | Log K |
|-------|-------|
| $Fe^{3+}$ | 25.70 |
| $Cu^{2+}$ | 18.80 |
| $Ni^{2+}$ | 18.56 |
| $Zn^{2+}$ | 16.50 |
| $Co^{2+}$ | 16.21 |
| $Fe^{2+}$ | 14.30 |
| $Mn^{2+}$ | 13.56 |
| $Ca^{2+}$ | 10.70 |
| $Mg^{2+}$ | 8.69 |

stability constants will be chelated first, followed in order of highest stability constant until the chelating agent is used.

Chelating agents are used to control the reactions of trace metals in foods to principally prevent discoloration, such as occurs in potatoes when iron reacts with phenolic compounds in the presence of oxygen. They are also used with antioxidants to complex trace metals, thus preventing the metal from acting as a catalyst in oxidative reactions. Application examples are the use of:

- Phosphates in soft drinks to chelate heavy metal ions that interfere with carbonation.
- EDTA in mayonnaise to eliminate the oxidative activities of trace metals and protect flavor.
- Sodium acid pyrophosphate to prevent discoloration in potatoes.
- Sodium hexametaphosphate to sequester calcium ions and permit the solubilization of alginates.

Examples of chelating agents include calcium disodium EDTA, disodium dihydrogen EDTA, tetrasodium pyrophosphate, citric acid, monoisopropyl citrate, phosphoric acid, and monoglyceride phosphate.

## COLORS

Colors are usually designated artificial or natural, which indicates that they are, respectively, synthetically manufactured or obtained from natural sources. Synthetic color additives "certified" by the Food and Drug Administration are designated FD&C (Food, Drug, and Cosmetic) and are traditionally termed primary colors. These colors are shades of red, yellow, blue, and green. Secondary colors are blends of certified primary colors, with or without diluents. Table 3 lists the physical and chemical properties of the certified food colors.

Those acceptable food colors not designated "certified" are designated "approved" and consist of natural organic and synthetic inorganic colorants used in certain applications (see Table 4). Colors are available in powders, liquids, granules, pastes, and other forms. Colorant determination includes desired hue, water solubility, and stability. The following groupings are used to illustrate different color groups.

**Table 3** Physical and Chemical Properties of Certified Food Colors

| FDA Name (Chemical Class) | Hue Range | Tinctorial Strength | Stability to: | | | Compatibility with Food Components |
| | | | Light | Oxidation | pH Change | |
| --- | --- | --- | --- | --- | --- | --- |
| FD&C Red #3 (erythrosine) | Bluish-pink | Very good | Fair | Fair | Poor | Poor |
| FD&C Red #40 | Yellowish-red | Very good | Very good | Fair | Good | Very good |
| FD&C Yellow #5 (tartrazine) | Lemon yellow | Good | Moderate | Fair | Good | Moderate |
| FD&C Yellow #6 (sunset yellow FCF) | Reddish | Good | Moderate | Fair | Good | Moderate |
| FD&C Green #3 (fast green FCF) | Bluish-green | Excellent | Fair | Poor | Good | Good |
| FD&C Blue #1 (brilliant blue FCF) | Greenish-blue | Excellent | Fair | Poor | Good (unstable in alkali) | Good |
| FD&C Blue #2 (indigotine) | Deep blue | Poor | Very poor | Poor | Poor | Very poor |

*Source:* Food Colors (National Academy of Sciences, 1971).

**Table 4** Physical and Chemical Properties of Some Noncertified Colors

| FDA Name (Chemical Class) | Hue Range | Tinctorial Strength | Stability to: | | | | Compatibility with Food Components |
|---|---|---|---|---|---|---|---|
| | | | Light | Oxidation | pH Change | Microbial Attack | |
| Annatto extract (carotenoid) | Yellow to peach | Good | Moderate | Very good | Very good | — | Very good |
| Beets, dehydrated (anthocyanin) | Bluish-red <pH 6 Brown > pH 6 | Good | Good | Excellent | Good | Good | Excellent |
| Caramel | Yellowish to tan | Fair | Good | Good | Good | Fair | Good |
| Beta-carotene (carotenoid) | Yellow to orange | Good | Fair | Poor | Good | Poor | Good |
| Cochineal extract; carmine | Orange red to wine red | Fair | Good | Good | Poor | Poor | Good |
| Grape skin extract | Red-blue-green | Poor | Poor | Poor | Poor | Fair | Good |
| Turmeric extract and oleoresin | Bright yellow to greenish-yellow | Good | Poor | Moderate | Poor | — | Good |

*Source:* Food Colors (National Academy of Sciences, 1971).

## Artificial Coloring

Water-soluble colors are designated as FD&C, followed by the color name and number designation, for example, FD&C Blue #2. They have a corresponding common name, for example, indigotine. The colors vary in hue, solubility, and other properties, which relates to the intended application. The water-soluble colors include FD&C Blue #1, Blue #2, Green #3, Red #40, Yellow #5, and Yellow #6.

Water-insoluble colors are termed FD&C aluminum lakes. Lakes are prepared by the absorption of a certified dye on an insoluble substrate, aluminum hydroxide, and as such include the standard colors. Lakes are used to color dry ingredients, increase stability, and reduce color migration. Lakes can be used to color foods with a high oil or fat content, in dry mixes and coated candies, and for other purposes.

## Natural Coloring

Natural colors are usually extracted from botanical sources and often contain several pigments and, as such, are not used as direct replacements for FD&C colors. The colors have low tinctorial strength due to a low quantity of pigment present and thus are used at higher levels than FD&C colors. These colors generally have poor stability in that their color and rate of degradation are affected by pH, temperature, and other conditions. Some natural colorings are:

- Annatto—The pigment bixin found in the coating of the annatto seed. The color hue ranges from yellow to reddish-orange.
- Turmeric—Contains curcumin obtained from turmeric root. The color hue varies from greenish-yellow to yellow-orange.
- Paprika—Produces a red to red-orange color.
- Beet—Produced from red beets and has a deep reddish-purple color.

## CORN SWEETENERS

Corn sweeteners are the products made by using mild conversion techniques to produce starch hydrolysates, that is, dextrin, maltodextrin, and corn-syrup solids. These carbohydrates vary according to sugar composition, which accounts for their differences in properties. They are classified in terms of dextrose equivalent (DE), a standard that expresses the level of reducing sugar calculated as dextrose (see Table 5). Complete conversion of cornstarch yields dextrose, also termed corn sugar, which has a DE of 92. When the reaction is stopped at the intermediate stage, corn syrup, consisting of dextrose, maltose, and polysaccharides, is obtained. It can be manufactured to have the desired DE based on desired properties. A common distribution is 28 to 38 DE (low conversion) to 58 to 68 DE (high conversion). Partial hydrolysis of cornstarch commercially produces maltodextrins of DE ranging from 13 to 22 and dextrin, with a DE range of 7 to 12. In high-fructose corn syrup, a fraction of the dextrose has been converted enzymatically to fructose, thus making a sweeter syrup.

## EMULSIFIERS

Emulsifiers are products that function to reduce the surface tension between two immiscible phases at their interface, allowing them to become miscible. The interface can be between two liquids, a liquid and a gas, or a liquid and a solid. Most emulsions involve water and oil or fat as the two immiscible phases, one being dispersed as finite globules in the other. The liquid as globules is referred to as the dispersed or internal phase, while the medium in which they are suspended is the continuous or external phase. There are two types of emulsions depending on the composition of the phases. In an oil-in-water emulsion such as milk and mayonnaise, the water is the external phase and the oil is the internal phase. In a water-in-oil emulsion such as butter, the oil is the external phase and the water is the internal phase. By use of the proper emulsifier, the two phases will mix and separation is prevented or delayed.

The emulsifier consists of a hydrocarbon chain which has affinity for fats and oils (lipophilic group) and a polar group which is attracted to

**Table 5** Comparison of Properties Relative to Dextrose Equivalent (DE)

| DE | Product | Solubility | Sweetness | Viscosity | Bodying Agent |
|----|---------|------------|-----------|-----------|---------------|
| 0 | Starch | 4 | 4 | 1 | 1 |
| 6–20 | Maltrodextrin | 3 | 3 | 2 | 2 |
| 20–58 | Corn Syrup | 2 | 2 | 3 | 3 |
| 100 | Corn Sugar (Dextrose) | 1 | 1 | 4 | 4 |

Range: 1 = greatest; 4 = least.

water or aqueous solutions (hydrophilic group). The emulsifier tends to concentrate at the interface between two immiscible liquids, with the hydrophilic portion in the water and the lipophilic portion in the oil. In such fashion, the surface properties are altered by the orientation of the emulsifiers at the interface which reduces the resistance of the two substances to combine. They are frequently used as blends for obtaining the most stable emulsion system. Emulsifiers have the following major functions:

- Complexing—Reaction with starch in bakery products which retards the crystallization of the starch, thus retarding the firming of the crumb which is associated with staling.
- Dispersing—The reduction of interfacial tension which creates an intimate mixture of two liquids that normally are immiscible, an example being oil-in-water emulsions such as salad dressing.
- Crystallization control—Control of crystallization in sugar and fat systems, i.e., chocolate, where it allows for brighter initial gloss and prevention of solidified fat on the surface.
- Wetting—Allows the surface to be more attracted to water, such as powders, i.e., coffee whitener, in which the addition of surfactant aids the dispersion of the powder in the liquid without lumping on the surface.
- Lubricating—Functions as a lubricant, such as in caramels, by reducing their tendency to stick to cutting knives, wrappers, and teeth.

Emulsifiers are also classified according to their solubility, being hydrophilic (water-loving) or lipophilic (oil-loving). The HLB, or hydrophilic/lipophilic balance, is a measure of the emulsifier's affinity to oil or water. The HLB range is 0 to 20, where 0 indicates completely lipophilic and 20 indicates completely hydrophilic. The behavior of emulsifiers in water according to the HLB range is shown in Table 6.

Table 7 lists the FDA names of food-grade emulsifiers and the Title 21 Code of Federal Regulations reference number for each.

## FATS AND OILS

Fats and oils belong to a group of substances classified as lipids which consist of the higher fatty acids and compounds associated with them. Lipids are characterized by their solubility in fat solvents, their insolubility in water, and their greasy feel. Fats and oils are of similar

**Table 6** Behavior of Emulsifiers in Water According to HLB Range

| Behavior When Added to Water | HLB Range |
|---|---|
| Insoluble | 1–4 |
| Poor solubility | 3–6 |
| Milky dispersion after vigorous agitation | 6–8 |
| Stable milky dispersion | 8–10 |
| Translucent-to-clear dispersion | 10–13 |
| Clear solution | 13+ |

HLB = Hydrophilic/lipophilic balance

**Table 7** FDA Names of Food-Grade Emulsifiers and the Title 21 Code of Federal Regulations References

| Emulsifier | Reference No. |
|---|---|
| Acteylated monoglycerides | 172.828 |
| Calcium stearyl-2-lactylate | 172.844 |
| Diacetyl tartaric acid esters of mono- and diglycerides | 184.1101 |
| Dioctyl sodium sulfosuccinate | 172.810 |
| Ethoxylated mono- and diglycerides | 172.834 |
| Fatty acids | 172.860 |
| Glyceryl-lacto esters of fatty acids | 172.852 |
| Hydroxylated lecithin | 172.814 |
| Lactylated fatty acid esters of glycerol and propylene glycol | 172.850 |
| Lactylic esters of fatty acids | 172.848 |
| Lecithin | 184.1400 |
| Mono- and diglycerides | 184.1505 |
| Monosodium phosphate derivatives of mono- and diglycerides | 184.1521 |
| Polyglycerol esters of fatty acids | 172.854 |
| Polyoxyethylene (20) sorbitan tristearate | 172.838 |
| Polyoxyethylene (20) sorbitan monostearate | 172.836 |
| Polyoxyethylene (20) sorbitan monooleate | 172.840 |
| Propylene glycol mono- and diesters of fats and fatty acids | 172.856 |
| Sodium lauryl sulfate | 172.822 |
| Sodium stearoyl-2-lactylate | 172.846 |
| Sodium stearyl fumarate | 172.826 |
| Sorbitan monostearate | 172.842 |
| Succinylated monoglycerides | 172.830 |
| Succistearin (stearoyl propylene glycol hydrogen succinate) | 172.765 |

chemical structure but differ physically in that at ordinary temperatures, fats are solids and oils are liquids. They are complex mixtures of predominantly mixed triglycerides, which are the compounds formed by combining one molecule of glycerol with three molecules of fatty acids. The fatty acids may be the same, two different fatty acids, or all different. Fatty acids are composed of a chain of carbon with hydrogen atoms, terminating in a carboxyl group. Fatty acids contain carbon chain lengths ranging from 4 to 24 and are identified according to the number of carbon atoms and whether they are saturated or unsaturated. Saturated fatty acids contain only single-bond carbon linkages and cannot accept additional hydrogen; unsaturated fatty acids have one or more double bonds and thus fewer hydrogen atoms and can accept hydrogen. Mono-unsaturated indicates that hydrogen can be accepted at one double-bond site; polyunsaturated indicates that hydrogen can be accepted at more than one double-bond site. The most highly unsaturated fats are oils, while fats of low unsaturation tend to be solids at room temperature. Hydrogenation (chemical addition of hydrogen to the double bond of unsaturated fatty acids) of a fat makes it firmer and more plastic, raises the melting point, and slows the development of rancidity by reducing the rate of reaction with oxygen. These fats are termed hydrogenated or partially hydrogenated oils. Fats and oils are composed of varying percentages of fatty acids which account for their respective properties. Some useful properties to consider in determining a suitable fat or oil include:

- Iodine value—An expression of the degree of unsaturation, which can serve as a guide in evaluating fat stability.
- Melting point—The temperature at which a solid changes to a liquid when heated.
- Solid fat index—A number indicating the proportion of solid to liquid present in the fat at a given temperature, which will reflect the consistency.

Fats and oils are used by themselves or as components of a food. Some examples of nomenclature are:

- Shortenings—Usually solid fats instead of oils used in baked goods to impart tenderness, soft crumb, etc.
- Spreads—Usually butter or margarine, which contains 80 percent or more fat.
- Salad oils—Oils which include olive, corn, cottonseed, soybean, sunflower.

- Cooking fat—Any edible fat or oil.
- Frying fat—A bland-flavored fat or oil of high smoking temperature to allow for heating to 400°F without smoking.
- Confectionary fat—A fat that is hard at room temperature and soft at body temperature, such as hydrogenated coconut oil or cacao butter.

Table 8 lists fats and oils with their principal component fatty acids.

## FLAVORS

Flavors are classified into the major groups of spices, natural flavors, and artificial flavors. A spice is an aromatic vegetable substance in a whole, broken, or ground form which is used as a seasoning. Natural flavors are flavor constituents derived from plant or animal sources. Artificial flavors are flavorings containing all or some portion of nonnatural materials.

Materials that can be used for flavorings can be grouped as follows: spices and herbs; essential oils and their extracts; fruits and fruit juices;

**Table 8** Fats and Oils and their Principal Component Fatty Acids

| Fats and Oils | Approximate Percentage of the Principal Fatty Acids |
|---|---|
| Butter Fat | 32 Oleic, 25 Palmitic |
| Cocoa Butter | 37 Oleic, 35 Stearic, 25 Palmitic |
| Coconut | 48 Lauric, 18 Myristic |
| Corn | 58 Linoleic, 26 Oleic |
| Cottonseed | 54 Linoleic, 24 Palmitic |
| Lard | 46 Oleic, 23 Palmitic |
| Olive | 68 Oleic, 14 Palmitic, 12 Linoleic |
| Palm | 47 Palmitic, 38 Oleic |
| Palm Kernel | 50 Lauric, 16 Myristic |
| Peanut | 46 Oleic, 31 Linoleic |
| Rapeseed | 31 Oleic, 23 Erucic, 19 Linoleic |
| Safflower | 75 Linoleic |
| Sesame | 43 Linoleic, 42 Oleic |
| Sorghum | 52 Linoleic, 31 Oleic |
| Soybean | 50 Linoleic, 25 Oleic |
| Sunflower | 68 Linoleic, 20 Oleic |
| Tallow, beef | 44 Oleic, 35 Palmitic |
| Tallow, mutton | 43 Oleic, 30 Stearic |

and aliphatic, aromatic, and terpene compounds. Spices and herbs consist of dried plant products that exhibit flavor and aroma. They are derived from true aromatic vegetable substances from which the volatile and flavoring principles have not been removed. Essential oils and their extracts are odorous oils obtained from plant material and have the major odor that is characteristic of that material. Most have poor water solubility and most contain terpenes (hydrocarbons of formula $C_{10}H_{16}$ and their oxygenated derivatives $C_{10}H_{16}O$ or $C_{10}H_{18}O$) which contribute to the poor water solubility as well as possibly contributing to the off-flavor. Examples are essential oils of bitter almond, anise, and clove. Terpeneless oils are extensions of concentrated essential oils in which the unwanted terpenes are removed. These oils are usually more concentrated and of increased stability and water solubility. Common oils in the terpeneless form are citrus oil, spearmint, and peppermint. Fruit and fruit juices are natural flavorings obtained from fruits. Whole, crushed, or pureed fruit may be used, but, more commonly, the juice or concentrate is used. Fruit extracts are made by extraction with a water-alcohol mixture. Aliphatic, aromatic, and terpene compounds refer to synthetic chemicals and isolates from natural materials. This classification encompasses the largest group of flavoring materials.

Flavors can be quite complex and the number of available flavors is extensive. Flavor is that property of a substance that causes a sensation of taste. Four basic tastes are perceived by taste buds on the tongue: sweet, salty, sour, and bitter. The flavors used are natural, artificial, or combinations and exist in liquid or dry form. General flavor types available include fruit, dairy, meat, vegetable, beverage, and liquor.

## FLOUR

Flour, also referred to as white flour, wheat flour, and plain flour, is the food prepared by grinding and bolting cleaned wheat other than durum wheat and red durum wheat (Code of Federal Regulations). Flour from other sources is available, identified according to its grain source. The properties of wheat flour vary according to the type of wheat, milling procedures, and treatment applied after milling.

Flour milling involves the separation of the endosperm, which is about 83 percent of the kernel, from other parts of the kernel, that is, bran and germ. The processing involves tempering, grinding, and sifting the large chunks of endosperm or "middlings" to yield the flour which, in the United States, will represent about 72 percent of the wheat kernel. By processing, size classification is achieved as flour streams. The streams include:

1. Straight flour—All the flour that can be milled from a wheat blend, or 72 percent of the wheat kernel which equates to 100 percent separation.
2. Long patent flour—90 to 95 percent separation.
3. Medium patent flour—80 to 90 percent separation.
4. Short patent flour—70 to 80 percent separation.
5. Short family or first patent flour—60 to 70 percent separation.
6. Extra short or fancy patent flour—40 to 60 percent separation.
7. Clears—Portion of straight flour remaining after removal of patent streams.

Flour properties depend upon the type of wheat, which is classified as hard or soft. Hard wheats are high in protein and the resulting flours have a high protein content and form a tenacious, elastic gluten with good gas-retaining properties and high water absorption capacity which makes it suitable for yeast-leavened bread. Soft wheats are low in protein and the resulting flour has poor gas-retaining properties and low water absorption capacity which makes it suitable for chemically leavened cakes and pastries.

The protein content of the flour is important because it forms the protein complex termed *gluten* when water and flour are kneaded together. The gluten formed accounts for the mixing and dough-handling characteristics as well as the formation of the framework of the baked product.

Different flours are used for different purposes. Some of these include:

- Bread flour, which generally contains in excess of 10.5 percent protein and is obtained from straight or long patent flours, has high absorption and good mixing tolerance.
- Cake flour, which generally contains less than 10 percent protein and is generally short patent flours, is low in absorption, and has short mixing time and tolerance.
- All-purpose (family) flour, which is intermediate between bread and cake flour.
- Pastry flour, which is obtained from soft wheat and can be straight or clear flour grades because color is not an essential requirement.
- Cracker flour, which generally contains 9 to 10.5 percent protein obtained from long patent or straight flours, is of low absorption, and has short mixing requirements.

Flours can be modified by various treatments to alter characteristics such as color, nutritional value, and baking qualities. Some of these modified flours include:

- Enriched flour—Flour that has been enriched by the inclusion of vitamins and minerals.
- Bromated flour—Potassium bromate has been added for improvement of baking qualities.
- Phosphated flour—Monocalcium phosphate has been added for improvement of baking qualities.
- Bleached flour—Flour in which the yellow carotenoid pigment has been converted to a nearly colorless product.

Other specific grain flours are obtained, with the term *flour* referring to that degree of grinding and sifting which results in a fine, powdery substance. The grain flours in Exhibit 1 are designated according to the grain from which they are obtained and include corn, rye, and durum flours.

## GUMS

Gums, or hydrocolloids, are polysaccharides that function as water-control agents in increasing viscosity (resistance to flow) or forming

---

**Exhibit 1** Cereal Flours Listed in the Code of Federal Regulations Part 137

| | |
|---|---|
| Flour | Whole durum flour |
| Bromated flour | White corn meal |
| Enriched bromated flour | Bolted white corn meal |
| Enriched flour | Enriched corn meals |
| Instantized flours | Degerminated white corn meal |
| Phosphated flour | Self-rising white corn meal |
| Self-rising flour | Yellow corn meal |
| Enriched self-rising flour | Bolted yellow corn meal |
| Cracked wheat | Degerminated yellow corn meal |
| Crushed wheat | Self-rising yellow corn meal |
| Whole wheat flour | Farina |
| Bromated whole wheat flour | Enriched farina |
| White corn flour | Semolina |
| Yellow corn flour | Enriched rice |
| Durum flour | |

gels. Gums are classified by source according to the following principal groupings: plant exudates, which include arabic, tragacanth, karaya, ghatti; seaweed extracts, which include agar, alginates, carrageenan, furcelleran; plant seed gums, which include guar, locust bean, tamarind, psyllium, quince; plant extracts, which include pectin and arabinogalactan; fermentation gums, which include xanthan gum, gellan gum, and dextran; and cellulose derivatives, which include carboxymethyl cellulose, hydroxypropylmethyl cellulose, microcrystalline cellulose. Gum derivatives include propylene glycol alginate and low-methoxy pectin. While starches and gelatin function as water-control agents, they are not included in this grouping.

The selection of a gum is based on the desired function and food application. By thickening or gelling the water, gums perform numerous roles such as stabilizers, film formers, binders, suspending agents, whipping agents, coating agents, and crystallization inhibitors. The gums perform these functions by themselves or in combination with other gums. Food properties considered in selecting a gum include pH, shelf stability, ingredient compatibility, texture, processing requirements, and ultimate consumer method of preparation. The differentiating properties of gums include viscosity, compatibility, pH stability, gel-forming capabilities, temperature stability, flow properties, and solubility. Within the same family of gums, there may be differences relative to salt type which will have an effect on its functional characteristics, that is, solubility, dispersibility, gel-forming capabilities, flow properties, and stability. Tables 9 and 10 list, respectively, comparative and relative properties of gums.

## PRESERVATIVES

Preservatives are antimicrobial agents. The preservatives most widely used are the benzoates (sodium benzoate), sorbates (sorbic acid and potassium sorbate), and the propionates (sodium or calcium propionates), which are organic acids or their salts (see Table 11). The activity of preservatives is due to the undissociated form of the molecule and thus pH is a major factor in their effectiveness. Increasing the acidity of foods is a method of controlling the growth of microorganisms. The survival and proliferation of microorganisms depends in part upon the pH of the food. Foods with a pH below 4.6 are considered acidic, and many bacteria will not proliferate in acidic foods. Acidulants are used to reduce the pH and thus provide a means of controlling microorganism growth. Acidulants used include acetic

**Table 9** Comparative Properties of Gums

| Gums | Cold Water Solubility | Hot Water Solubility | Gel Former | Acid Stability |
|---|---|---|---|---|
| Agar | No | Yes | Yes | Between pH 4.5–9.0 |
| Alginate, sodium | Yes | Yes | Yes | Gels at pH 3.5 depending on calcium content |
| Arabic (Acacia) | Yes | Yes | No | pH 4–10 |
| Carboxymethyl cellulose | Yes | Yes | No | Best between pH 7–9, below pH 5 get reduction in viscosity |
| Carrageenan | No, except lambda & sodium salts | Yes | Yes, except lambda | Solution undergoes hydrolysis at acid pH (3.5); gel is stable |
| Furcelleren | No | Yes | Yes | Heating below pH 5 causes hydrolysis and gel degradation |
| Gelatin | No—swells | Yes | Yes | Stable, gradual decline in gel strength with acidification |
| Gellan gum, high acyl | No | Yes | Yes | Stable in acid pHs |
| Gellan gum, low acyl | No | Yes | Yes | Stable in acid pHs |
| Ghatti | Yes | Yes | No | Opt. viscosity at pH 8, drops on both sides |

| | | | | |
|---|---|---|---|---|
| Guar | Yes | Yes | No | Between pH 3.5–10.5, gradual decline with acidification |
| Hydroxypropyl cellulose | Yes | No, insoluble above 45°C | No | Between pH 3–10, opt. pH 6–8 |
| Hydroxypropyl methylcellulose | Yes | No | Yes, at elevated temp. depending on type | pH 3–11 |
| Karaya | Yes, swells | Some | No | Viscosity decreased by acids or electrolytes |
| Locust bean gum | Swells, requires heat | Yes | No | Between pH 5–8, at higher or lower values get considerable variation |
| Low-methoxy pectin | Yes, depends on methoxy content | Yes | Yes | Form gels between pH 2.5–6.5 depending on system |
| Methylcellulose | Yes | No | Yes, at elevated temp | pH 3–11 |
| Microcrystalline cellulose | Insoluble, dispersible | Insoluble, dispersible | No | Insoluble, resistant |
| Pectin | Yes | Yes | Yes | Gel below pH 3.6 |
| Psyllium | Yes | Yes | Yes, at high conc. | Between pH 2–10 |
| Quince | Yes | Yes | No | Between pH 4–10 |
| Tragacanth | Yes, swells | Yes | No | Between pH 4–6 |
| Xanthan gum | Yes | Yes | No | Between pH 2–12 |

**Table 10** Relative Properties of Gelling Gums

| Property | Agar | Carregeenan | Furcelleran | Gelatin | Gellan gum, high acyl | Gellan gum, low acyl | LM Pectin | Pectin | Sodium Alginate | Xanthan Gum/Locust Bean Gum |
|---|---|---|---|---|---|---|---|---|---|---|
| Solubility | >90°C | Kappa 50–60°C Iota 50–60°C Lambda room temp. | 70–80°C | 70°C | 70–80°C, swells initially | 75–90°C | Room temp. | Room temp. | Room temp. | 70°C |
| Gelling Temperature | Set 32–39°C Melt >85–90°C | Kappa, Iota—set & melt temp. are 10°C apart, vary with solute; Lambda: nongelling | Set—40°C Melt—50°C, vary with solute | Set 20°C Melt 30°C | Set/Melt—at high temp. with some ion affect | Set—room temp. to 50°C, depending on ion concentration. Melt >80°C, depending on ion concentration | At some fixed temp. depending on system | Set 50–99°C Melt 70–100°C + depends on type of pectin, soluble solids, etc. | Room temp. | 49–55°C |
| Mechanism | Cooling | K, Ca ions | Cooling | Cooling | Ca ions, cools | Ca ions, cooling | Ca ions | Cooling; pH/ soluble solids | Ca ions | Cooling |
| Reversibility | Thermo-reversible | Thermo-reversible | Thermo-reversible | Thermo-reversible | Thermo-reversible | Non-reversible | Thermo-reversible | Nonreversible; will melt but not reset | Non-reversible | Thermo-reversible |

| | | | | | | | | |
|---|---|---|---|---|---|---|---|---|
| Transparency | Turbid | Kappa—transparent in pot. form. Iota—transparent in both forms | Pot. form—transparent. Ca form—turbid | Transparent | Opaque | Clear | Turbid, transparent | Clear |
| Texture | Brittle | Kappa—brittle Iota—elastic | Brittle/elastic | Elastic | Elastic, cohesive | Firm, brittle | Brittle | Brittle/elastic |
| Syneresis | Yes | Kappa—yes Iota—no | Yes | No | Some | Some | Yes | Some |

| | | |
|---|---|---|
| Clear | Turbid |
| Elastic | Elastic |
| Some | No |

**Table 11** Preservatives

| Preservative Category | Activity | Use Level |
|---|---|---|
| Benzoates (sodium benzoate) | Yeasts, molds & bacteria, but usually not recommended for bacterial control because of restricted use level & lower activity at higher pH; best activity at pH 2.5–4.0 | 0.01–0.10% |
| Sorbates (sorbic acid, potassium sorbate) | Yeasts & molds, least activity against bacteria & on a selective basis; best activity up to pH 6.5 | 0.03–0.10% |
| Parabens | Yeasts & molds, less active against bacteria, especially gram-negative; effective up to pH 8.0 | 0.10% range |
| Propionates (propionic acid, Ca, Na propionate) | Molds, slightly antibacterial action except against "rope"; effective up to pH 6.0 | 0.20–0.50% |

acid, adipic acid, citric acid, fumaric acid, lactic acid, and phosphoric acid. Greater effectiveness is achieved in acidic systems. Sorbic acid and potassium sorbate have the best activity up to pH 6.5, calcium and sodium propionate up to pH 5.0, and sodium benzoate up to pH 4.5.

Sorbic acid and potassium sorbate are effective against yeast and mold inhibition with little activity against bacteria. Common uses are in cheese, sausage, and baked goods not including yeast-raised goods. Sorbic acid has low solubility in water which increases with increasing temperature, while potassium sorbate is readily soluble in water. Potassium sorbate has the same antimycotic properties as sorbic acid and on an equivalent weight basis has 74 percent of the activity of sorbic acid. Thus, higher concentrations are required to obtain the same yeast and mold-inhibiting effects (4 parts potassium sorbate equal 3 parts sorbic acid).

Calcium and sodium propionate are effective against molds and have slight antibacterial action and little action on yeasts. Application areas include baked goods and processed cheese. Because they have little action against yeasts, they can be used in yeast-baked goods and are the most common preservative in baked goods.

Sodium benzoate is effective against yeasts and slightly effective against bacteria and molds. The most effective range is pH 2.5 to 4.0

with a maximum pH of 4.5. It is used in acidulated beverages, jams, jellies, and relishes.

Parabens, which are esters of para-hydroxybenzoic acid, are related to benzoic acid but are effective over a wider pH range. They are active against yeasts and molds and are used in baked goods and beverages.

## SPICES

Spices consist of dried plant products that exhibit flavor and aroma. Spices are obtained from vegetable substances from which none of the volatile or other flavoring substances have been removed. Spices are grouped into (a) tropical spices such as pepper and cloves; (b) herbs, such as sage and rosemary; (c) spicy seeds such as mustard and anise; and (d) dehydrated aromatic vegetables such as onion and garlic.

Spices in the ground form have an increased surface area and consequently the oil glands are ruptured, causing the evaporation of the essential oil and loss of aroma. Spice flavor is also obtained by use of an extract or essential oil which carries the spice aroma in a concentrated form. These oils are volatile and as such do not contain the nonvolatile constituents. The compound containing both the essential oil and the nonvolatile constituents is commercially known as oleoresins, which contain all the odorous and flavor principles of the spice. Oleoresins offer flavor uniformity, stability, freedom from bacteria, and flavor concentration.

Spices are used predominantly in prepared meats, luncheon meats, sauces, salads, soups, and dressings. Other important users are bakers, pickle packers, condiment manufacturers, and the canning industry.

A list of spices and other natural seasonings and flavorings, found in Title 21, Section 182.10, of the Code of Federal Regulations, is in Table 12.

## STARCH

Starch, consisting of repeating glucose units, is separated into the polysaccharides amylose and amylopectin. Amylose consists of straight chains containing 200 to 2,100 glucose units, while amylopectin consists of branched chains containing 20 to 25 glucose units each. A visible difference is that amylose is more soluble and less viscous than amylopectin and facilitates gel formation. Starches vary in their amylose content.

**Table 12** Spices and Other Natural Seasonings and Flavorings

| Common Name | Botanical Name of Plant Source |
| --- | --- |
| Alfalfa herb and seed | *Medicago sativa* L. |
| Allspice | *Pimenta officinalis* Lindl. |
| Ambrette seed | *Hibiscus abelmoschus* L. |
| Angelica | *Angelica archangelica* L. or other spp. of *Angelica.* |
| Angelica root | Do. |
| Angelica seed | Do. |
| Angostura (cusparia bark) | *Galipea officinalis* Hancock. |
| Anise | *Pimpinella anisum* L. |
| Anise, star | *Illicium verum* Hook. f. |
| Balm (lemon balm) | *Melissa officinalis* L. |
| Basil, bush | *Ocimum minimum* L. |
| Basil, sweet | *Ocimum basilicum* L. |
| Bay | *Laurus nobilis* L. |
| Calendula | *Calendula officinalis* L. |
| Camomile (chamomile), English or Roman | *Anthemis nobilis* L. |
| Camomile (chamomile), German or Hungarian | *Matricaria chamomilla* L. |
| Capers | *Capparis spinosa* L. |
| Capsicum | *Capsicum frutescens* L. or *Capsicum annum* L. |
| Caraway | *Carum carvi* L. |
| Caraway, black (black cumin) | *Nigella sativa* L. |
| Cardamon (cardamon) | *Elettaria cardamomum* Maton. |
| Cassia, Chinese | *Cinnamomum cassia* Blume. |
| Cassia, Padang or Batavia | *Cinnamomum burmanni* Blume. |
| Cassia, Saigon | *Cinnamomum loureirii* Nees. |
| Cayenne papper | *Capsicum frutescens* L. or *Capsicum annuum* L. |
| Celery seed | *Apium graveolens* L. |
| Chervil | *Anthriscus cerefolium* (L.) Hoffm. |
| Chives | *Allium schoenprasum* L. |
| Cinnamon, Ceylon | *Cinnamomum zeylanicum* Nees. |
| Cinnamon, Chinese | *Cinnamomum cassia* Blume. |
| Cinnamon, Saigon | *Cinnamomum loureirii* Nees. |
| Clary (clary sage) | *Salvia sclarea* L. |
| Clover | *Trifolium* spp. |
| Coriander | *Coriandrum sativum* L. |
| Cumin (cummin) | *Cuminum cyminum* L. |
| Cumin, black (black caraway) | *Nigella sativa* L. |
| Elder flowers | *Sambucus canadensis* L. |
| Fennel, common | *Foeniculum vulgare* Mill. |
| Fennel, sweet (finocchio, Florence fennel) | *Foeniculum vulgare* Mill. var. duice (DC.) Alex. |

**Table 12** *continued*

| Common Name | Botanical Name of Plant Source |
| --- | --- |
| Fenugreek | *Trigonella foenum-graecum* L. |
| Galanga (galangal) | *Alpinia officinarum* Hance. |
| Geranium | *Pelargonium* spp. |
| Ginger | *Zingiber officinale* Rosc. |
| Grains of paradise | *Amomum melegueta* Rosc. |
| Horehound (hoarhound) | *Marrubium vulgare* L. |
| Horseradish | *Armoracia lapathifolia* Gilib. |
| Hyssop | *Hyssopus officinalis* L. |
| Lavender | *Lavandula officinalis* Chaix. |
| Linden flowers | *Tilia* spp. |
| Mace | *Myristica fragrans* Houtt. |
| Marigold, pot | *Calendula officinalis* L. |
| Marjoram, pot | *Majorana onites* (L.) Benth. |
| Marjoram, sweet | *Majorana hortensis* Moench. |
| Mustard, black or brown | *Brassica nigra* (L.) Koch. |
| Mustard, brown | *Brassica juncea* (L.) Coss. |
| Mustard, white or yellow | *Brassica hirta* Moench. |
| Nutmeg | *Myristica fragrans* Houtt. |
| Oregano (oreganum, Mexican oregano, Mexican sage, origan) | *Lippia* spp. |
| Paprika | *Capsicum annuum* L. |
| Parsley | *Petroselinum crispum* (Mill.) Mansf. |
| Pepper, black | *Piper nigrum* L. |
| Pepper, cayenne | *Capsicum frutescens* L. or *Capsicum annuum* L. |
| Pepper, red | Do. |
| Pepper, white | *Piper nigrum* L. |
| Peppermint | *Mentha piperita* L. |
| Poppy seed | *Papayer somniferum* L. |
| Pot marigold | *Calendula officinalis* L. |
| Pot marjoram | *Majorana onites* (L.) Benth. |
| Rosemary | *Rosmarinus officinalis* L. |
| Saffron | *Crocus sativus* L. |
| Sage | *Salvia officinalis* L. |
| Sage, Greek | *Salvia triloba* L. |
| Savory, summer | *Satureia hortensis* L. (Satureja). |
| Savory, winter | *Satureia montana* L. (Satureja). |
| Sesame | *Sesamum indicum* L. |
| Spearmint | *Mentha spicata* L. |
| Star anise | *Illicium verum* Hook. f. |
| Tarragon | *Artemisia dracunculus* L. |

*continues*

**Table 12** *continued*

| Common Name | Botanical Name of Plant Source |
| --- | --- |
| Thyme | *Thymus vulgaris* L. |
| Thyme, wild or creeping | *Thymus serpyllum* L. |
| Turmeric | *Curcuma longa* L. |
| Vanilla | *Vanilla planifolia* Andr. or *Vanilla tahitensis* J. W. Moore. |
| Zedoary | *Curcuma zedoaria* Rosc. |

*Note:* Spices and other natural seasonings and flavorings are listed in Title 21, Section 182.10 of the Code of Federal Regulations.

---

Waxy starches, so termed because the cut endosperm resembles hard, opaque wax, contain mostly amylopectin, while ordinary cornstarch consists of about 24 percent amylose and 76 percent amylopectin. The waxy starches form thick, clear pastes but gel only at high concentrations such as 30 percent, while 4 to 5 percent cornstarch will form a gel. The paste viscosity of the waxy maize starch remains the same hot or cold. High-amylose starches contain 50 to 70 percent amylose and have unique properties for functioning as film formers, oxygen and fat barriers, quick-setting stable gels, and binders.

Starch is not soluble in cold water, but forms a suspension. Upon heating the suspension to the gelatinization temperature (60 to 70°C), the starch granules suddenly swell, the opaque suspension slowly becomes translucent upon continued heating, and the viscosity increases to the thickness of a boiled starch paste. Most starches require heating to 90°C in order to obtain a firm gel upon cooling, when the viscosity increases and may form a gel depending on the type of starch. During storage of a starch paste or gel, the molecules become less soluble and tend to aggregate and partially crystallize; the change is termed *retrogradation*, which is the opposite of gelatinization. The starch gel shrinks and some of the liquid separates from the gel. Retrogradation does not occur in waxy starches because they do not contain amylose.

Starches can be modified (they are then termed *modified starches*) by chemical modification or cross-linking, to provide desired properties not found in natural starch. By the production of cross-links or bridges from one starch molecule to another, starch can be made more resistant to hydrolysis, thus preventing the loss of viscosity. The resistance of starch to shear or mixing is directly proportional to the degree of cross-linking. Other properties obtained include viscosity control, freeze-thaw stability, heat resistance, and acid resistance. These starches find application in sauces, gravies, pie filling, frozen foods, and other products where specific properties are required.

Pregelantinized starch is a starch processed to swell to some degree in cold water unlike regular starch, which requires heating. The most common method involves heating a starch paste to its gelatinization temperature, drying on a drum dryer, and grinding the dried starch to a powder. Upon reconstitution with water, the pregelatinized starch has less thickening power and tendency to gel than pastes of the parent starch. This starch is used in applications requiring more rapid hydration or room temperature preparation, such as instant desserts, puddings, and soups. Based on abundance and cost, cornstarch is the most commonly used. Other available starches include grain sorghum, rice, wheat, potato, tapioca, arrowroot, and waxy varieties.

## SWEETENERS

Sweeteners can be classified as natural or artificial. The natural sweeteners are carbohydrates consisting of molecules of carbon, hydrogen, and oxygen. The simplest form of carbohydrate is the monosaccharide or simple sugar and includes glucose (dextrose), fructose (levulose), and galactose, which are six-carbon (hexose) sugars.

The combination of two monosaccharides forms a disaccharide sugar, which can also be formed by the breakdown of longer-chain carbohydrates termed *polysaccharides*. The following combinations of monosaccharides form the respective disaccharides: glucose + frucose = sucrose; glucose + galactose = lactose; glucose + glucose = maltose.

**Table 13** Nutritive Sweeteners Sweetness Relative to Sucrose

| | |
|---|---|
| Levulose | 173 |
| Fructose | 150 |
| Honey (dry basis) | 100–150 |
| Invert sugar | 130 |
| **Sucrose** | 100 |
| High fructose corn syrup | 70–80 |
| Dextrose (glucose) | 70 |
| Corn syrup | 40–70 |
| Maltose | 30–35 |
| Galactose | 32 |
| Lactose | 15–20 |

A trisaccharide consists of three monosaccharides, such as raffinose which consists of galactose, glucose, and fructose. A tetrasaccharide such as stachyose consists of four monosaccharides. These more complicated sugars are not digestible so they are not used as sweeteners. A polysaccharide is a longer-chained carbohydrate which exists in digestible and nondigestible forms. The digestible forms are starch, a polymer of glucose units from which corn sweeteners are obtained, and glycogen, a polymer of glucose which is the carbohydrate reserve of animals. The nondigestible form includes cellulose, lignin, and gums such as pectin and algin.

Polyhydric alcohols (polyols) in foods consist of glycerine, sorbitol, mannitol, propylene glycol, and xylitol. When used at low levels, the taste of the polyols is of minor consequence; but when used as a major ingredient, such as in sugarless chewing gum, the polyol is the major source of sweetness.

Sucrose is the most widely used natural sweetener and is usually the reference relative to sweetness, taste profile, and cost. Corn sweeteners, fructose, and high intensity sweeteners are other widely used sweeteners.

Sweeteners are also grouped as nutritive and non-nutritive. Nutritive sweeteners include sucrose, fructose, dextrose, lactose, maltose, honey, high fructose corn syrups, and polyols. Non-nutritive sweeteners (artificial sweeteners) include saccharine, aspartame, acesulfame-K, and sucralose.

The relative sweetness of sweeteners relative to sucrose can vary according to concentration, temperature, etc. (Table 13).

Polyols (polyhydric alcohols, sugar alcohols) are produced by hydrogenating the corresponding reducing sugars, for example, Sorbitol—hydrogenated from glucose. They provide the bulk and texture of sucrose but have less kilocalories/gram. Polyols include sorbitol, mannitol, xylitol, erythritol, lacticol, maltitol, and isomalt (Table 14).

**Table 14** Polyols Sweetness Relative to Sucrose

| | |
|---|---|
| **Sucrose** | 100 |
| Xylitol | 100 |
| Maltitol | 90 |
| Erythritol | 60–70 |
| Sorbitol | 60 |
| Isomalt (palatinit) | 50–60 |
| Mannitol | 50 |
| Lacticol | 30–40 |

## High Intensity Sweeteners

High intensity sweeteners are products intended to imitate the taste of sucrose and function as non-nutritive replacements; they provide basically 0 calories. Saccharin, discovered in 1878, has the longest history of food use.

The properties of high intensity sweeteners vary according to source, relative sucrose sweetness, taste profile, solubility, stability to temperature and pH, synergies, and applications. These sweeteners are used singly or in combination to maximize the beneficial properties of each (Table 15).

## VITAMINS

Vitamins are organic compounds that are essential for normal body growth and maintenance. They are classified into groups: fat-soluble vitamins—vitamins A, D, E, and K; and water-soluble vitamins—vitamins B and C. Vitamins are measured in very low concentrations, such as 1 to 100 mg. Through biochemical action, they perform various functions in such processes as cell growth, normal digestion, manufacture of red blood cells, and absorption of calcium and phosphorus. Inadequate vitamin intake can be the result of food deficiency, increased vitamin requirements, and increased vitamin loss. The vitamins of determined importance include: A (retinol), $B_1$ (thiamine), $B_2$ (riboflavin), $B_5$ (pantothenic acid), $B_6$ (pyridoxine), $B_{12}$ (cyanocobalamin), C (ascorbic acid), $D_2$ (calciferol), E (tocopherol), K, niacin, folic acid, and biotin.

Table 16 describes the functions and provides sources for fat- and water-soluble vitamins.

**Table 15** High Intensity Sweetness Relative to Sucrose

| | |
|---|---|
| Sucralose | 600 |
| Saccharin | 300 |
| Acesulfame-K | 200 |
| Aspartame | 200 |
| **Sucrose** | 100 |

*Note:* Relative sweetness can vary according to concentration and temperature.

**Table 16** Vitamin Functions and Sources

| Vitamins | Function | Sources |
| --- | --- | --- |
| *Fat Soluble* | | |
| Vitamin A (retinol) | Necessary for cell growth, healthy skin; prevents night blindness | Green & yellow fruits & vegetables; eggs, butter, cheese |
| Vitamin $D_2$ (calciferol) | Necessary for bone & teeth growth; deficiency causes rickets | Fish, liver, oil, vit. D milk, sunshine |
| Vitamin E (tocopherol) | Functions as antioxidant, preventing the oxidation of unsaturated fatty acids & protecting vitamins such as vit. A | Legumes, meat, eggs, whole grains |
| Vitamin K | Essential for blood clotting | Green leafy vegetables, liver, soybeans |
| *Water Soluble* | | |
| Vitamin $B_1$ (thiamine) | Necessary for growth, fertility, lactation; deficiency causes beriberi | Pork, fish, cereal, beans, peas |
| Vitamin $B_2$ (riboflavin) | Necessary for growth; acts as coenzyme | Milk, cheese, eggs, poultry |
| Vitamin $B_6$ (pyridoxine hydrochloride) | Functions as coenzyme; involved in utilization of protein | Meat, corn, lima beans |
| Vitamin $B_{12}$ (cobalamin) | Necessary for normal functioning of cells | Meat, liver, dry milk |
| Niacin | Necessary for healthy cells & tissues; prevents pellagra | Meat, liver, enriched bread |
| Pantothenic acid | Necessary for several bodily functions | Whole grain cereal, meat, fish |
| Biotin | Necessary in metabolism | Peanuts, beans, eggs, meat |
| Folic acid | Necessary in metabolism; helps manufacture red blood cells | Leafy green vegetables, yeast, liver |
| Vitamin C (ascorbic acid) | Essential for healthy bones & teeth; contributes to resistance to infection; deficiency causes scurvy | Fresh fruit & vegetables |

# PART III

# Additives/Substances for Use in Foods

## Listed under Title 21 of the Code of Federal Regulations

*Note:* Refer to Code of Federal Regulations, Title 21, for complete information on usage.

# PART 73—LISTING OF COLOR ADDITIVES EXEMPT FROM CERTIFICATION

## Subpart A—Foods

| | |
|---|---|
| 73.1 | Diluents in color additive mixtures for food use exempt from certification |
| 73.30 | Annatto extract |
| 73.35 | Astaxanthin |
| 73.40 | Dehydrated beets (beet powder) |
| 73.50 | Ultramarine blue |
| 73.75 | Canthaxanthin |
| 73.85 | Caramel |
| 73.90 | Beta-apo-8'-carotenal |
| 73.95 | Beta-carotene |
| 73.100 | Cochineal extract, carmine |
| 73.140 | Toasted partially defatted cooked cottonseed flour |
| 73.160 | Ferrous gluconate |
| 73.165 | Ferrous lactate |
| 73.169 | Grape color extract |
| 73.170 | Grape skin extract (enocianina) |
| 73.200 | Synthetic iron oxide |
| 73.250 | Fruit juice |
| 73.260 | Vegetable juice |
| 73.275 | Dried algae meal |
| 73.295 | Tagetes (Aztec marigold) meal and extract |
| 73.300 | Carrot oil |
| 73.315 | Corn endosperm oil |
| 73.340 | Paprika |
| 73.345 | Paprika oleoresin |
| 73.450 | Riboflavin |
| 73.500 | Saffron |
| 73.575 | Titanium dioxide |
| 73.600 | Turmeric |
| 73.615 | Turmeric oleoresin |

# PART 74—LISTING OF COLOR ADDITIVES SUBJECT TO CERTIFICATION

## Subpart A—Foods

| | |
|---|---|
| 74.101 | FD&C Blue No. 1 |
| 74.102 | FD&C Blue No. 2 |

| 74.203 | FD&C Green No. 3 |
|--------|------------------|
| 74.250 | Orange B |
| 74.302 | Citrus Red No. 2 |
| 74.303 | FD&C Red No. 3 |
| 74.340 | FD&C Red No. 40 |
| 74.705 | FD&C Yellow No. 5 |
| 74.706 | FD&C Yellow No. 6 |

## PART 172—FOOD ADDITIVES PERMITTED FOR DIRECT ADDITION TO FOOD FOR HUMAN CONSUMPTION

### Subpart A—General Provisions

172.5    General provisions for direct food additives: Food additives may be used under conditions of good manufacturing practice.

### Subpart B—Food Preservatives

| 172.105 | Anoxomer |
|---------|----------|
| 172.110 | BHA |
| 172.115 | BHT |
| 172.120 | Calcium disodium EDTA |
| 172.130 | Dehydroacetic acid |
| 172.133 | Dimethyl dicarbonate |
| 172.135 | Disodium EDTA |
| 172.140 | Ethoxyquin |
| 172.145 | Heptylparaben |
| 172.150 | 4-Hydroxymethyl-2,6-di-tert-butylphenol |
| 172.155 | Natamycin (pimaricin) |
| 172.160 | Potassium nitrate |
| 172.165 | Quaternary ammonium chloride combination |
| 172.170 | Sodium nitrate |
| 172.175 | Sodium nitrite |
| 172.177 | Sodium nitrite used in processing smoked chub |
| 172.180 | Stannous chloride |
| 172.185 | TBHQ |
| 172.190 | THBP |

### Subpart C—Coatings, Films, and Related Substances

172.210    Coatings on fresh citrus fruit

172.515    Synthetic flavoring substances and adjuvants
172.520    Cocoa with dioctyl sodium sulfosuccinate for manufacturing
172.530    Disodium guanylate
172.535    Disodium inosinate
172.540    DL-Alanine
172.560    Modified hop extract
172.575    Quinine
172.580    Safrole-free extract of sassafras
172.585    Sugar beet extract flavor base
172.590    Yeast-malt sprout extract

## Subpart G—Gums, Chewing Gum Bases, and Related Substances

172.610    Arabinogalactan
172.615    Chewing gum base
172.620    Carrageenan
172.623    Carrageenan with polysorbate 80
172.626    Salts of carrageenan
172.655    Furcelleran
172.660    Salts of furcelleran
172.665    Gellan gum
172.695    Xanthan gum

## Subpart H—Other Specific Usage Additives

172.710    Adjuvants for pesticide use dilutions
172.712    1,3-Butylene glycol
172.715    Calcium lignosulfonate
172.720    Calcium lactobionate
172.723    Epoxidized soybean oil
172.725    Gibberellic acid and its potassium salt
172.730    Potassium bromate
172.735    Glycerol ester of wood rosin
172.755    Stearyl monoglyceridyl citrate
172.765    Succistearin (stearoyl propylene glycol hydrogen succinate)
172.770    Ethylene oxide polymer
172.775    Methacrylic acid–divinylbenzene copolymer

## Subpart I—Multipurpose Additives

| | |
|---|---|
| 172.800 | Acesulfame potassium |
| 172.802 | Acetone peroxides |
| 172.804 | Aspartame |
| 172.806 | Azodicarbonamide |
| 172.808 | Copolymer condensates of ethylene oxide and propylene oxide |
| 172.809 | Curdlan |
| 172.810 | Dioctyl sodium sulfosuccinate |
| 172.811 | Glyceryl tristearate |
| 172.812 | Glycine |
| 172.814 | Hydroxylated lecithin |
| 172.816 | Methyl glucoside–coconut oil ester |
| 172.818 | Oxystearin |
| 172.820 | Polyethylene glycol (mean molecular weight 200–9500) |
| 172.822 | Sodium lauryl sulfate |
| 172.824 | Sodium mono- and dimethyl naphthalene sulfonates |
| 172.826 | Sodium stearyl fumarate |
| 172.828 | Acetylated monoglycerides |
| 172.830 | Succinylated monoglycerides |
| 172.831 | Sucralose |
| 172.832 | Monoglyceride citrate |
| 172.833 | Sucrose acetate isobutyrate (SAIB) |
| 172.834 | Ethoxylated mono- and diglycerides |
| 172.836 | Polysorbate 60 |
| 172.838 | Polysorbate 65 |
| 172.840 | Polysorbate 80 |
| 172.841 | Polydextrose |
| 172.842 | Sorbitan monostearate |
| 172.844 | Calcium stearoyl-2-lactylate |
| 172.846 | Sodium stearoyl lactylate |
| 172.848 | Lactylic esters of fatty acids |
| 172.850 | Lactylated fatty acid esters of glycerol and propylene glycol |
| 172.852 | Glyceryl-lacto esters of fatty acids |
| 172.854 | Polyglycerol esters of fatty acids |
| 172.856 | Propylene glycol mono- and diesters of fats and fatty acids |
| 172.858 | Propylene glycol alginate |
| 172.859 | Sucrose fatty acid esters |
| 172.860 | Fatty acids |
| 172.861 | Cocoa butter substitute from coconut oil, palm kernel oil, or both oils |

| | |
|---|---|
| 172.862 | Oleic acid derived from tall oil fatty acids |
| 172.863 | Salts of fatty acids |
| 172.864 | Synthetic fatty alcohols |
| 172.866 | Synthetic glycerin produced by the hydrogenolysis of carbohydrates |
| 172.867 | Olestra |
| 172.868 | Ethyl cellulose |
| 172.870 | Hydroxypropyl cellulose |
| 172.872 | Methyl ethyl cellulose |
| 172.874 | Hydroxypropyl methylcellulose |
| 172.876 | Castor oil |
| 172.878 | White mineral oil |
| 172.880 | Petrolatum |
| 172.882 | Synthetic isoparaffinic petroleum hydrocarbons |
| 172.884 | Odorless light petroleum hydrocarbons |
| 172.886 | Petroleum wax |
| 172.888 | Synthetic petroleum wax |
| 172.890 | Rice bran wax |
| 172.892 | Food starch–modified |
| 172.894 | Modified cottonseed products intended for human consumption |
| 172.896 | Dried yeasts |
| 172.898 | Bakers yeast glycan |

## PART 182—SUBSTANCES GENERALLY RECOGNIZED AS SAFE

While not including all substances that are generally recognized as safe (GRAS) for their intended use, such as salt, pepper, and vinegar, this listing includes substances that, when used for that purpose indicated and in accordance with good manufacturing practice, are regarded as GRAS for such uses.

### Subpart A—General Provisions

| | |
|---|---|
| 182.1 | Substances that are generally recognized as safe |
| 182.10 | Spices and other natural seasonings and flavorings |
| 182.20 | Essential oils, oleoresins (solvent-free), and natural extractives (including distillates) |
| 182.40 | Natural extractives (solvent-free) used in conjunction with spices, seasonings, and flavorings |

## Subpart B—Multiple Purpose GRAS Food Substances

## Subpart C—Anticaking Agents

182.2729    Sodium calcium aluminosilicate, hydrated
182.2906    Tricalcium silicate


## Subpart D—Chemical Preservatives

182.3013    Ascorbic acid
182.3041    Erythorbic acid
182.3089    Sorbic acid
182.3109    Thiodipropionic acid
182.3149    Ascorbyl palmitate
182.3169    Butylated hydroxyanisole
182.3173    Butylated hydroxytoluene
182.3189    Calcium ascorbate
182.3225    Calcium sorbate
182.3280    Dilauryl thiodipropionate
182.3616    Potassium bisulfite
182.3637    Potassium metabisulfite
182.3640    Potassium sorbate
182.3731    Sodium ascorbate
182.3739    Sodium bisulfite
182.3766    Sodium metabisulfite
182.3795    Sodium sorbate
182.3798    Sodium sulfite
182.3862    Sulfur dioxide
182.3890    Tocopherols


## Subpart G—Sequestrants

182.6085    Sodium acid phosphate
182.6197    Calcium diacetate
182.6203    Calcium hexametaphosphate
182.6215    Monobasic calcium phosphate
182.6285    Dipotassium phosphate
182.6290    Disodium phosphate
182.6757    Sodium gluconate
182.6760    Sodium hexametaphosphate
182.6769    Sodium metaphosphate
182.6778    Sodium phosphate
182.6787    Sodium pyrophosphate
182.6789    Tetra sodium pyrophosphate
182.6810    Sodium tripolyphosphate

## Subpart H—Stabilizers

182.7255     Chondrus extract

## Subpart I—Nutrients

182.8013     Ascorbic acid
182.8159     Biotin
182.8217     Calcium phosphate
182.8223     Calcium pyrophosphate
182.8250     Choline bitartrate
182.8252     Choline chloride
182.8778     Sodium phosphate
182.8890     Tocopherols
182.8892     α-Tocopherol acetate
182.8985     Zinc chloride
182.8988     Zinc gluconate
182.8991     Zinc oxide
182.8994     Zinc stearate
182.8997     Zinc sulfate

## PART 184—DIRECT FOOD SUBSTANCES AFFIRMED AS GENERALLY RECOGNIZED AS SAFE

The list of direct human food ingredients are considered GRAS under the conditions prescribed and shall be used in accordance with current good manufacturing practice.

## Subpart A—General Provisions

184.1     Substances added directly to human food affirmed as generally recognized as safe (GRAS)

## Subpart B—Listing of Specific Substances Affirmed as GRAS

184.1005     Acetic acid
184.1007     Aconitic acid
184.1009     Adipic acid

184.1011   Alginic acid
184.1012   α-Amylase enzyme preparation from *Bacillus stearothermo-philus*
184.1021   Benzoic acid
184.1024   Bromelain
184.1025   Caprylic acid
184.1027   Mixed carbohydrase and protease enzyme product
184.1033   Citric acid
184.1034   Catalase (bovine liver)
184.1061   Lactic acid
184.1063   Enzyme-modified lecithin
184.1065   Linoleic acid
184.1069   Malic acid
184.1077   Potassium acid tartrate
184.1081   Propionic acid
184.1090   Stearic acid
184.1091   Succinic acid
184.1095   Sulfuric acid
184.1097   Tannic acid
184.1099   Tartaric acid
184.1101   Diacetyl tartaric acid esters of mono- and diglycerides
184.1115   Agar-agar
184.1120   Brown algae
184.1121   Red algae
184.1133   Ammonium alginate
184.1135   Ammonium bicarbonate
184.1137   Ammonium carbonate
184.1138   Ammonium chloride
184.1139   Ammonium hydroxide
184.1140   Ammonium citrate, dibasic
184.1141a  Ammonium phosphate, monobasic
184.1141b  Ammonium phosphate, dibasic
184.1143   Ammonium sulfate
184.1148   Bacterially derived carbohydrase enzyme preparation
184.1150   Bacterially derived protease enzyme preparation
184.1155   Bentonite
184.1157   Benzoyl peroxide
184.1165   n-Butane and iso-butane
184.1185   Calcium acetate
184.1187   Calcium alginate
184.1191   Calcium carbonate
184.1193   Calcium chloride

184.1195    Calcium citrate
184.1199    Calcium gluconate
184.1201    Calcium glycerophosphate
184.1205    Calcium hydroxide
184.1206    Calcium iodate
184.1207    Calcium lactate
184.1210    Calcium oxide
184.1212    Calcium pantothenate
184.1221    Calcium propionate
184.1229    Calcium stearate
184.1230    Calcium sulfate
184.1240    Carbon dioxide
184.1245    Beta-carotene
184.1250    Cellulase enzyme preparation derived from *Trichoderma longibrachiatum*
184.1257    Clove and its derivatives
184.1259    Cocoa butter substitute
184.1260    Copper gluconate
184.1261    Copper sulfate
184.1262    Corn silk and corn silk extract
184.1265    Cuprous iodide
184.1271    L-Cysteine
184.1272    L-Cysteine monohydrochloride
184.1277    Dextrin
184.1278    Diacetyl
184.1282    Dill and its derivatives
184.1287    Enzyme-modified fats
184.1293    Ethyl alcohol
184.1295    Ethyl formate
184.1296    Ferric ammonium citrate
184.1297    Ferric chloride
184.1298    Ferric citrate
184.1301    Ferric phosphate
184.1304    Ferric pyrophosphate
184.1307    Ferric sulfate
184.1307a   Ferrous ascorbate
184.1307b   Ferrous carbonate
184.1307c   Ferrous citrate
184.1307d   Ferrous fumarate
184.1308    Ferrous gluconate
184.1311    Ferrous lactate
184.1315    Ferrous sulfate

184.1316   Ficin
184.1317   Garlic and its derivatives
184.1318   Glucono delta-lactone
184.1321   Corn gluten
184.1322   Wheat gluten
184.1323   Glyceryl monooleate
184.1324   Glyceryl monostearate
184.1328   Glyceryl behenate
184.1329   Glyceryl palmitostearate
184.1330   Acacia (gum arabic)
184.1333   Gum ghatti
184.1339   Guar gum
184.1343   Locust (carob) bean gum
184.1349   Karaya gum (sterculia gum)
184.1351   Gum tragacanth
184.1355   Helium
184.1366   Hydrogen peroxide
184.1370   Inositol
184.1372   Insoluble glucose isomerase enzyme preparations
184.1375   Iron, elemental
184.1386   Isopropyl citrate
184.1387   Lactase enzyme preparation from *Candida pseudotro-picalis*
184.1388   Lactase enzyme preparation from *Kluyveromyces lactis*
184.1400   Lecithin
184.1408   Licorice and licorice derivatives
184.1409   Ground limestone
184.1415   Animal lipase
184.1420   Lipase enzyme preparation derived from *Rhizopus niveus*
184.1425   Magnesium carbonate
184.1426   Magnesium chloride
184.1428   Magnesium hydroxide
184.1431   Magnesium oxide
184.1434   Magnesium phosphate
184.1440   Magnesium stearate
184.1443   Magnesium sulfate
184.1443a  Malt
184.1444   Maltodextrin
184.1445   Malt syrup (malt extract)
184.1446   Manganese chloride
184.1449   Manganese citrate
184.1452   Manganese gluconate

| 184.1461 | Manganese sulfate |
| 184.1472 | Menhaden oil |
| 184.1490 | Methylparaben |
| 184.1498 | Microparticulated protein product |
| 184.1505 | Mono- and diglycerides |
| 184.1521 | Monosodium phosphate derivatives of mono- and diglycerides |
| 184.1530 | Niacin |
| 184.1535 | Niacinamide |
| 184.1537 | Nickel |
| 184.1538 | Nisin preparation |
| 184.1540 | Nitrogen |
| 184.1545 | Nitrous oxide |
| 184.1553 | Peptones |
| 184.1555 | Rapeseed oil |
| 184.1560 | Ox bile extract |
| 184.1563 | Ozone |
| 184.1583 | Pancreatin |
| 184.1585 | Papain |
| 184.1588 | Pectins |
| 184.1595 | Pepsin |
| 184.1610 | Potassium alginate |
| 184.1613 | Potassium bicarbonate |
| 184.1619 | Potassium carbonate |
| 184.1622 | Potassium chloride |
| 184.1625 | Potassium citrate |
| 184.1631 | Potassium hydroxide |
| 184.1634 | Potassium iodide |
| 184.1635 | Potassium iodate |
| 184.1639 | Potassium lactate |
| 184.1643 | Potassium sulfate |
| 184.1655 | Propane |
| 184.1660 | Propyl gallate |
| 184.1666 | Propylene glycol |
| 184.1670 | Propylparaben |
| 184.1676 | Pyridoxine hydrochloride |
| 184.1685 | Rennet (animal-derived) and chymosin preparation (fermentation-derived) |
| 184.1695 | Riboflavin |
| 184.1697 | Riboflavin-5'-phosphate (sodium) |
| 184.1698 | Rue |
| 184.1699 | Oil of rue |

184.1702   Sheanut oil
184.1721   Sodium acetate
184.1724   Sodium alginate
184.1733   Sodium benzoate
184.1736   Sodium bicarbonate
184.1742   Sodium carbonate
184.1751   Sodium citrate
184.1754   Sodium diacetate
184.1763   Sodium hydroxide
184.1764   Sodium hypophosphite
184.1768   Sodium lactate
184.1769a  Sodium metasilicate
184.1784   Sodium propionate
184.1792   Sodium sesquicarbonate
184.1801   Sodium tartrate
184.1804   Sodium potassium tartrate
184.1807   Sodium thiosulfate
184.1835   Sorbitol
184.1845   Stannous chloride (anhydrous and dihydrated)
184.1848   Starter distillate
184.1851   Stearyl citrate
184.1854   Sucrose
184.1857   Corn sugar
184.1859   Invert sugar
184.1865   Corn syrup
184.1866   High fructose corn syrup
184.1875   Thiamine hydrochloride
184.1878   Thiamine mononitrate
184.1890   $\alpha$-Tocopherols
184.1901   Triacetin
184.1903   Tributyrin
184.1911   Triethyl citrate
184.1914   Trypsin
184.1923   Urea
184.1924   Urease enzyme preparation from *Lactobacillus fermentum*
184.1930   Vitamin A
184.1945   Vitamin B$_{12}$
184.1950   Vitamin D
184.1973   Beeswax (yellow and white)
184.1976   Candelilla wax
184.1978   Carnauba wax
184.1979   Whey

184.1979a  Reduced lactose whey
184.1979b  Reduced minerals whey
184.1979c  Whey protein concentrate
184.1983    Bakers yeast extract
184.1984    Zein
184.1985    Aminopeptidase enzyme preparation derived from lactococcus lactis

## PART 186—INDIRECT FOOD SUBSTANCES AFFIRMED AS GENERALLY RECOGNIZED AS SAFE

The list of indirect human food ingredients are considered GRAS for the purpose and conditions prescribed, providing they comply with the purity specifications or are of a purity suitable for their intended use.

### Subpart A—General Provisions

186.1       Substances added indirectly to human food affirmed as generally recognized as safe (GRAS)

### Subpart B—Specific Substances Affirmed as GRAS

186.1093  Sulfamic acid
186.1256  Clay (kaolin)
186.1275  Dextrans
186.1300  Ferric oxide
186.1316  Formic acid
186.1374  Iron oxides
186.1551  Hydrogenated fish oil
186.1555  Japan wax
186.1557  Tall oil
186.1673  Pulp
186.1750  Sodium chlorite
186.1756  Sodium formate
186.1770  Sodium oleate
186.1771  Sodium palmitate
186.1797  Sodium sulfate
186.1839  Sorbose

# PART IV

# Food Additives E Numbers in the European Union

## FOOD ADDITIVES E NUMBERS

Food Additives in the European Union permitted for use in foods according to directive as established by the European Scientific Committee for Food.


## E NUMBER LIST

| | |
|---|---|
| E100 | Curcumin |
| E101 | (i) Riboflavin, (ii) Riboflavin-5'-phosphate |
| E102 | Tartrazine |
| E104 | Quinoline Yellow |
| E110 | Sunset Yellow FCF, Orange Yellow S |
| E120 | Cochineal, Carminic acid, Carmines |
| E122 | Azorubine, Carmoisine |
| E123 | Amaranth |
| E124 | Ponceau 4R, Cochineal Red A |
| E127 | Erythrosine |
| E128 | Red 2G |
| E129 | Allura Red AC |
| E131 | Patent Blue V |
| E132 | Indigotine, Indigo carmine |
| E133 | Brilliant Blue FCF |
| E140 | Chlorophylls and Chlorophyllins: (i) Chlorophylls, (ii) Chlorophyllins |
| E141 | Copper complexes of chlorophylls and chlorophyllins: (i) Copper complexes of chlorophylls, (ii) Copper complexes of chlorphyllins |
| E142 | Greens S |
| E150a | Plain caramel |
| E150b | Caustic sulphite caramel |
| E150c | Ammonia caramel |
| E150d | Sulphite ammonia caramel |
| E151 | Brilliant Black BN, Black PN |
| E153 | Vegetable carbon |
| E154 | Brown FK |
| E155 | Brown HT |
| E160a | Carotenes: (i) Mixed carotenes, (ii) Beta-carotene |
| E160b | Annatto, bixin, norbixin |
| E160c | Paprika extract, capsanthin, capsorubin |
| E160d | Lycopene |
| E160e | Beta-apo-8'-carotenal (C 30) |
| E160f | Ethyl ester of beta-apo-8'-carotenic acid (C 30) |
| E161b | Lutein |
| E161g | Canthaxanthin |
| E162 | Beetroot Red, betanin |

E163    Anthocyanins
E170    Calcium carbonates
E171    Titanium dioxide
E172    Iron oxides and hydroxides
E173    Aluminum
E174    Silver
E175    Gold
E180    Latolrubine BK
E200    Sorbic acid
E202    Potassium sorbate
E203    Calcium sorbate
E210    Benzoic acid
E211    Sodium benzoate
E212    Potassium benzoate
E213    Calcium benzoate
E214    Ethyl p-hydroxybenzoate
E215    Sodium ethyl p-hydroxybenzoate
E216    Propyl p-hydroxybenzoate
E217    Sodium propyl p-hydroxybenzoate
E218    Methyl p-hydroxybenzoate
E219    Sodium methyl p-hydroxybenzoate
E220    Sulphur dioxide
E221    Sodium sulphite
E222    Sodium hydrogen sulphite
E223    Sodium metabisulphite
E224    Potassium metabisulphite
E226    Calcium sulphite
E227    Calcium hydrogen sulphite
E228    Potassium hydrogen sulphite
E230    Biphenyl, diphenyl
E231    Orthophenyl phenol
E232    Sodium orthophenyl phenol
E233    Thiabendazole
E234    Nisin
E235    Natamycin
E239    Hexamethylene tetramine
E242    Dimethyl dicarbonate
E250    Sodium nitrite
E251    Sodium nitrate
E252    Potassium nitrate
E260    Acetic acid
E261    Potassium acetate
E262    Sodium acetates: (i) Sodium acetate, (ii) Sodium hydrogen acetate
        (sodium diacetate)
E263    Calcium acetate

E270    Lactic acid
E280    Propionic acid
E281    Sodium propionate
E282    Calcium propionate
E283    Potassium propionate
E284    Boric acid
E285    Sodium tetraborate (borax)
E249    Potassium nitrite
E290    Carbon dioxide
E296    Malic acid
E297    Fumaric acid
E300    Ascorbic acid
E301    Sodium ascorbate
E302    Calcium ascorbate
E304    Fatty acid esters of ascorbic acid: (i) Ascorbyl palmitate, (ii) Ascorbyl stearate
E306    Tocopherol-rich extract
E307    Alpha-tocopherol
E308    Gamma-tocopherol
E309    Delta-tocopherol
E310    Propyl gallate
E311    Octyl gallate
E312    Dodecyl gallate
E315    Erythorbic acid
E316    Sodium erythorbate
E320    Butylated hydroxyanisole (BHA)
E321    Butylated hydroxytoluene (BHT)
E322    Lecithins
E325    Sodium lactate
E326    Potassium lactate
E327    Calcium lactate
E330    Citric acid
E331    Sodium citrates: (i) Monosodium citrate, (ii) Disodium citrate, (iii) Trisodium citrate
E332    Potassium citrates: (i) Monopotassium citrate, (ii) Tripotassium citrate
E333    Calcium citrates: (i) Monocalcium citrate, (ii) Dicalcium ci-trate, (iii) Tricalcium citrate
E334    Tartaric acid (L(+)-)
E335    Sodium tartrates: (i) Monosodium tartrate, (ii) Disodium tartrate
E336    Potassium tartrates: (i) Monopotassium tartrate, (ii) Dipotassium tartrate
E337    Sodium potassium tartrate
E338    Phosphoric acid

E339    Sodium phosphates: (i) Monosodium phosphate, (ii) Disodium phosphate, (iii) Trisodium phosphate

E340    Potassium phosphates: (i) Monocalcium phosphate, (ii) Dipotassium phosphate, (iii) Tripotassium phosphate

E341    Calcium phosphates: (i) Monocalcium phosphate, (ii) Dicalcium phosphate, (iii) Tricalcium phosphate

E343    Magnesium phosphates: (i) Monomagnesium phosphate, (ii) Dimagnesium phosphate

E350    Sodium malates: (i) Sodium malate, (ii) Sodium hydrogen malate

E351    Potassium malate

E352    Calcium malates: (i) Calcium malate, (ii) Calcium hydrogen malate

E353    Metatartaric acid

E354    Calcium tartrate

E355    Adipic acid

E356    Sodium adipate

E357    Potassium adipate

E363    Succinic acid

E380    Triammonium citrate

E385    Calcium disodium ethylene diamine tetra-acetate (Calcium disodium EDTA)

E400    Alginic acid

E401    Sodium alginate

E402    Potassium alginate

E403    Ammonium alginate

E404    Calcium alginate

E405    Propan-1,2-diol alginate

E406    Agar

E407    Carrageenan

E407a   Processed eucheuma seaweed

E410    Locust bean gum

E412    Guar gum

E413    Tragacanth

E414    Acacia gum (gum arabic)

E415    Xanthan gum

E416    Karaya gum

E417    Tara gum

E418    Gellan gum

E420    Sorbitol: (i) Sorbitol, (ii) Sorbitol syrup

E421    Mannitol

E422    Glycerol

E425    Konjac: (i) Konjac gum, (ii) Konjac glucomannane

E431    Polyoxyethylene (40) stearate

E432    Polyoxyethylene sorbitan monolaurate (polysorbate 20)

E433    Polyoxyethylene sorbitan monooleate (polysorbate 80)

E434    Polyoxyethylene sorbitan monopalmitate (polysorbate 40)

| | |
|---|---|
| E435 | Polyoxyethylene sorbitan monostearate (polysorbate 60) |
| E436 | Polyoxyethylene sorbitan tristearate (polysorbate 65) |
| E440 | Pectins: (i) Pectin, (ii) Amidated pectin |
| E442 | Ammonium phosphatides |
| E444 | Sucrose acetate isobutyrate |
| E445 | Glycerol esters of wood rosins |
| E450 | Diphosphates: (i) Disodium diphosphate, (ii) Trisodium diphosphate, (iii) Tetrasodium diphosphate, (iv) Dipotassium diphosphate, (v) Tetrapotassium diphosphate, (vi) Dicalcium diphosphate, (vii) Calcium dihydrogen disphosphate |
| E451 | Triphosphates: (i) Pentasodium triphosphate, (ii) Pentapotassium triphosphate |
| E452 | Polyphosphates: (i) Sodium polyphosphates, (ii) Potassium polyphosphates, (iii) Sodium calcium polyphosphate, (iv) Calcium polyphosphates |
| E459 | Beta-cyclodextrine |
| E460 | Cellulose: (i) Microcrystalline cellulose, (ii) Powdered cellulose |
| E461 | Methyl cellulose |
| E463 | Hydroxypropyl cellulose |
| E464 | Hydroxypropyl methyl cellulose |
| E465 | Ethyl methyl cellulose |
| E466 | Carboxy methyl cellulose, Sodium carboxy methyl cellulose |
| E468 | Crosslinked sodium carboxymethyl cellulose |
| E469 | Enzymically hydrolysed carboxy methyl cellulose |
| E470a | Sodium, potassium and calcium salts of fatty acids |
| E470b | Magnesium salts of fatty acids |
| E471 | Mono- and diglycerides of fatty acids |
| E472a | Acetic acid esters of mono- and diglycerides of fatty acids |
| E472b | Lactic acid esters of mono- and diglycerides of fatty acids |
| E472c | Citric acid esters of mono- and diglycerides of fatty acids |
| E472d | Tartaric acid esters of mono- and diglycerides of fatty acids |
| E472e | Mono- and diacetyl tartaric acid esters of mono- and diglycerides of fatty acids |
| E472f | Mixed acetic and tartaric acid esters of mono- and diglycerides of fatty acids |
| E473 | Sucrose esters of fatty acids |
| E474 | Sucroglycerides |
| E475 | Polyglycerol esters of fatty acids |
| E476 | Polyglycerol polyricinoleate |
| E477 | Propane-1,2-diol esters of fatty acids |
| E479b | Thermally oxidized soya bean oil interacted with mono- and diglycerides of fatty acids |
| E481 | Sodium stearoyl-2-lactylate |
| E482 | Calcium stearoyl-2-lactylate |
| E483 | Stearyl tartrate |

| | |
|---|---|
| E491 | Sorbitan monostearate |
| E492 | Sorbitan tristearate |
| E493 | Sorbitan monolaurate |
| E494 | Sorbitan monooleate |
| E495 | Sorbitan monopalmitate |
| E500 | Sodium carbonates: (i) Sodium carbonate, (ii) Sodium hydrogen carbonate, (iii) Sodium esquicarbonate |
| E501 | Potassium carbonates: (i) Potassium carbonate, (ii) Potassium hydrogen carbonate |
| E503 | Ammonium carbonates: (i) Ammonium carbonate, (ii) Ammonium hydrogen carbonate |
| E504 | Magnesium carbonates: (i) Magnesium carbonate, (ii) Magnesium hydroxide carbonate (syn. Magnesium hydrogen carbonate) |
| E507 | Hydrochloric acid |
| E508 | Potassium chloride |
| E509 | Calcium chloride |
| E511 | Magnesium chloride |
| E512 | Stannous chloride |
| E513 | Sulphuric acid |
| E514 | Sodium sulphates: (i) Sodium sulphate, (ii) Sodium hydrogen sulphate |
| E515 | Potassium sulphates: (i) Potassium sulphate, (ii) Potassium hydrogen sulphate |
| E516 | Calcium sulphate |
| E517 | Ammonium sulphate |
| E520 | Aluminum sulphate |
| E521 | Aluminum sodium sulphate |
| E522 | Aluminum potassium sulphate |
| E523 | Aluminum ammonium sulphate |
| E524 | Sodium hydroxide |
| E525 | Potassium hydroxide |
| E526 | Calcium hydroxide |
| E527 | Ammonium hydroxide |
| E528 | Magnesium hydroxide |
| E529 | Calcium oxide |
| E530 | Magnesium oxide |
| E535 | Sodium ferrocyanide |
| E536 | Potassium ferrocyanide |
| E538 | Calcium ferrocyanide |
| E541 | Sodium aluminum phosphate, acidic |
| E551 | Silicon dioxide |
| E552 | Calcium silicate |
| E553a | (i) Magnesium silicate, (ii) Magnesium trisilicate |
| E553b | Talc |
| E554 | Sodium aluminum silicate |

| | |
|---|---|
| E555 | Potassium aluminum silicate |
| E556 | Calcium aluminium sicicate |
| E558 | Bentonite |
| E559 | Aluminum silicate (Kaolin) |
| E570 | Fatty acids |
| E574 | Gluconic acid |
| E575 | Glucono-delta-lactone |
| E576 | Sodium gluconate |
| E577 | Potassium gluconate |
| E578 | Calcium gluconate |
| E579 | Ferrous gluconate |
| E585 | Ferrous lactate |
| E620 | Glutamic acid |
| E621 | Monosodium glutamate |
| E622 | Monopotassium glutamate |
| E623 | Calcium diglutamate |
| E624 | Monoammonium glutamate |
| E625 | Magnesium diglutamate |
| E626 | Guanylic acid |
| E627 | Disodium guanylate |
| E628 | Dipotassium guanylate |
| E629 | Calcium guanylate |
| E630 | Inosinic acid |
| E631 | Disodium inosinate |
| E632 | Dipotassium inosinate |
| E633 | Calcium inosinate |
| E634 | Calcium 5'-ribonucleotides |
| E635 | Disodium 5'-ribonucleotides |
| E640 | Glycine and its sodium salt |
| E900 | Dimethyl polysiloxane |
| E901 | Beeswax, white and yellow |
| E902 | Candelilla wax |
| E903 | Carnauba wax |
| E904 | Shellac |
| E905 | Microcystalline wax |
| E912 | Montanic acid esters |
| E914 | Oxidized polyethylene wax |
| E920 | L-Cysteine |
| E927b | Carbamide |
| E938 | Argon |
| E939 | Helium |
| E941 | Nitrogen |
| E942 | Nitrous oxide |
| E948 | Oxygen |
| E950 | Acesulfame K |

E951    Aspartame
E952    Cyclamic acid and its Na and Ca salts
E953    Isomalt
E954    Saccharin and its Na, K, and Ca salts
E957    Thaumatin
E959    Neohesperidine DC
E965    Maltitol: (i) Maltitol, (ii) Maltitol syrup
E966    Lactitol
E967    Xylitol
E999    Quillaia extract
E1103   Invertase
E1105   Lysozyme
E1200   Polydextrose
E1201   Polyvinylpyrrolidone
E1202   Polyvinylpolypyrrolidone
E1404   Oxidized starch
E1410   Monostarch phosphate
E1412   Distarch phosphate
E1413   Phosphated distarch phosphate
E1414   Acetylated distarch phosphate
E1420   Acetylated starch
E1422   Acetylated distarch adipate
E1440   Hydroxy propyl starch
E1442   Hydroxy propyl distarch phosphate
E1451   Acetylated oxidized starch
E1450   Starch sodium octenyl succinate
E1505   Triethyl citrate
E1518   Glyceryl triacetate (triacetin)

# PART V

# Bibliography

*Note:* The bibliography represents references used to create the dictionary and update its ingredient listing.

# BIBLIOGRAPHY

*Adipic Acid*. St. Louis, MO: Monsanto.

*Adipic Acid*. Product Bulletin. Dallas, TX: Celanese Chemical Co., Inc.

*Agar Agar*. Technical Bulletin. North Bergen, N.J.: Frutarom Incorporated.

Akoh, C. C. 1998. *Fat Replacers*. Food Technology 52(3): 47–53.

*Agave Nectar*. Technical Bulletin. City of Industry, CA: Western Commerce Corporation.

*Alberger Salt for Food Processing*. St. Clair, MI: Diamond Crystal Salt Co.

Altschul, A. *Protein Forum*. Decatur, IL: A. E. Staley Mfg. Co.

*Aluminum Sulfate*. Technical Bulletin. Morristown, NJ: Allied Chemical Company.

*Amaizo Corn Syrups*. Hammond, IN: American Maize-Products Co.

*Amaizo Food Specialty Starches*. 1976. Hammond, IN: American Maize-Products Co.

*Amaizo Fro-Dex® Corn Syrup Solids*. 1979. Hammond, IN: American Maize-Products Co.

American Dairy Products Institute. 1987. Select the Whey Product to Meet Your Ingredient Needs. *Food Technology* 41(10): 124–125.

Andres, C. 1988. Antioxidants—"Quality Protectors." *Food Processing* 46(2): 36–41.

Andres, C. 1984. Black Pepper—The Versatile Flavor. *Food Processing* 45(5): 84–86.

Andres, C. 1980. Corn—A Most Versatile Grain. *Food Processing* 41(5): 78–100.

Andres, C. 1982. Food Type, Operating Procedures, Food pH Enter into Color Decision. *Food Processing* 43(5): 52–53.

Andres, C. 1984. Gluten—The Versatile Ingredient. *Food Processing* 45(5): 80–82.

Andres, C. 1982. Mold/Yeast Inhibitor Grains FDA Approval for Cheese. *Food Processing* 43(11): 83.

Andres, C. 1975. Processors Guide to Gums—Part I. *Food Processing* 36(12): 35–36.

Andres, C. 1976. Stabilizers 2—Gums. *Food Processing* 37(1): 83–87.

Andres, C. 1984. Sulfites. *Food Processing* 46(3): 34.

Andres, C. 1976. Surfactants/Emulsifiers Improve Texture, Quality, Shelf Life of Foods. *Food Processing* 37(5): 63–68.

Andres, C. 1984. Three Factors Affecting Price/Supply. *Food Processing* 45(4): 77.

Andres, C. 1982. Yellow and Reddish-Orange Colors Have Good Stability and Solubility. *Food Processing* 43(5): 44–45.

*Annatto Colors.* Technical Bulletin. Milwaukee, WI: Chr. Hansen's Laboratory, Inc.

*Annatto Colors.* Technical Bulletin. Elkhart, IN: Miles Laboratories, Inc.

*Anti-caking and Conditioning Agents.* 1977. Edison, NJ: J. M. Huber Corp.

*The Apple. The Pectin.* Technical Bulletin. Neuenbürg, FR Germany: Herman Herbstreith KG.

*Aromatics.* Technical Bulletin. Los Angeles, CA: F. Ritter & Co.

*Ascorbyl Palmitate NF,FCC.* Product Data Bulletin. Nutley, NJ: Roche Chemical Division.

*Atlas Food Color Guide.* New York: H. Kohnstamm & Co., Inc.

*Atlas Products for Foods.* Wilmington, DE: ICI United States, Inc.

*Avicel® PH 101 in Grated and Shredded Cheese.* Technical Bulletin. Philadelphia: FMC Corp.

*Avicel® PH Microcrystalline Cellulose.* Technical Bulletin. Philadelphia: FMC Corp.

*Avicel® RC.* Technical Bulletin. Philadelphia: FMC Corp.

*Bacteria, Yeast & Molds.* 1972. Athens, GA: Cooperative Extension Service, University of Georgia.

*Barley.* Cannon Falls, MN: Minnesota Grain Pearling Company.

*Basic Sodium Aluminum Phosphate.* Technical Bulletin. Westport, CT: Stauffer Chemical Co.

*Benzoic Acid & Sodium Benzoate.* 1970. St. Louis, MO: Monsanto.

Best, D. 2000. Flaxseed—Not Just for Pets! *Food Processing* 61(6): 66–69.

Brandt, L. A. 1998. Function Follows Formulation. *Prepared Foods* 167(7): 61–63.

Brandt, L. A. 2000. The Whole Sugar-Free Scoop. *Prepared Foods* 169(3): 47–50.

*Bulgar Wheat.* Technical Bulletin. Fresno, CA: Sunnyland Mills.

*Calcium Lactate.* Gorinchem, Holland: C. V. Chemie Combinatie Amsterdam.

*Calcium Sulfate for Food and Pharmaceutical Uses.* Technical Bulletin. Chicago: United States Gypsum.

*Calcium Sulfate for the Baking Industry.* 1978. Technical Bulletin. Chicago: United States Gypsum.

*Cal Plus™.* Technical Bulletin. St. Louis, MO: Mallinckrodt, Inc.

*Cal Tac™.* Technical Bulletin. St. Louis, MO: Mallinckrodt, Inc.

*Caramel Coloring.* Cedar Rapids, IA: Corn Swseeteners.

The Carrageenan People. Technical Bulletin. Philadelphia, PA: FMC Corporation.

Cassidy, J. P. 1977. Phosphates in the Dairy Industry. *Food Product Development* 11(5): 83–85.

*Cerelose® Dextrose.* 1980. Englewood Cliffs, NJ: CPC International, Inc.

*Certified FD&C Colors.* 1979. Technical Bulletin. Cincinnati, OH: Hilton-Davies.

*Chemicals Used in Food Processing.* 1965. Washington, DC: National Academy of Sciences, National Research Council.

Chedd, G. 1974. The Search for Sweetness. *New Scientist* (May): 299–302.

*Chocolate Information.* Technical Bulletin. Hunt Valley, MD: McCormick & Co., Inc.

*The Choice is Clear.* Technical Bulletin. Minneapolis, MN: Novartis Nutrition.

*Citric Acid.* Technical Bulletin No. 78. New York: Pfizer Chemicals Division.

*Code of Federal Regulations, Title 21, Parts 1–99, Parts 100–169, Parts 170–199.* April 1, 2000. National Archive and Records Administration.

*A Comparison of the Properties of Pure Crystalline Fructose and Isomerized Corn Syrups.* Technical Bulletin. Nutley, NJ: Roche Chemical Division.

*Corn Bran-Regular.* Technical Bulletin. Paris, IL: Illinois Cereal Mills, Inc.

*Creative Cooking with Bulgur Wheat.* Danville, IL: Lauhoff Grain Company.

*Current FDA Status of Atlas® Surfactants and Polyols for Use in Foods.* Technical Bulletin. Wilmington, DE: ICI United States, Inc.

*CWS Fumaric Acid.* St. Louis, MO: Monsanto.

Damon, G. E. 1974. A Primer on Vitamins. *FDA Consumer* (May): 5–11.

*Dehydroacetic Acid.* 1980. Publication No. A-106A. Kingsport, TN: Eastman Chemical Products, Inc.

*Delvocid®.* Product Bulletin Del-03/82.10 Am.10. Charlotte, NC: G. B. Fermentation Industries, Inc.

Dietary Salt. 1980. *Food Technology* 34(1): 85–90.

*Distilled Monoglyceride Type S, Type S (V).* Tokyo, Japan: Riken Vitamin Oil Co., Ltd.

*Durkee Product Catalog.* 1979. Cleveland, OH: Durkee Industrial Foods Group.

*Durkee Shortening & Oil Glossary.* Cleveland, OH: Durkee Industrial Foods Group.

*Durkote GDL 135-50.* 1978. Technical Bulletin. Cleveland, OH: Durkee Industrial Foods Group.

Earl, E., and J. F. Rand. 1977. Foam Control in Potato Processing. *Food Engineering* 49(11): 71–73.

Eastman Chemical Co. 2000. Beverage Weighting Agent Recently Approved in the U.S. *Food Technology* 54(5): 152.

*Edifas®A Methyl Ethyl Cellulose.* 1974. Technical Bulletin. Wilmington, DE: ICI America, Inc.

*Edible Rennet Casein.* New Zealand: New Zealand Dairy Board.

*Eggs in Brief.* 1985. Park Ridge, IL: The American Egg Board.

*Eggs, Your Diet and Your Health.* Park Ridge, IL: National Commission on Egg Nutrition.

Emodi, A. 1978. Xylitol, Its Properties and Food Applications. *Food Technology* 32(1): 28–32.

Engineered Ingredients: Building Blocks for Food Design. 1975. *Food Engineering* 47(7): 46–47.

*Enocyanin.* Technical Bulletin. Plainview, NY: Burlington Bio-Medical & Scientific Corporation.

*Erythorbic Acid and Sodium Erythorbate in Foods.* 1975. New York, NY: Pfizer Chemicals Division.

*Erythritol.* Technical Bulletin. Hammond, IN: Cerestar U.S.A., Inc.

Esselen, W. B. 1975. Applications for Rosells as a Red Food Colorant. *Food Product Development* 9(8): 37–40.

*Essential Oils and Oleoresins.* Madison, WI: Quality Control Spice Co., Inc.

*"Fagopyrum sagittatum." Or "buckwheat."* 1980. Cannon Falls, MN: Minnesota Grain Pearling Company.

FDA Approves Acesulfame-K for Use in Foods and Beverages. 1988. *Food Engineering International* 13(8): 16.

*Flavor Enhancers and Nutritional Ingredients.* Technical Bulletin. Clifton, NJ: Yeast Products, Inc.

*Flour.* Manhattan, KS: American Institute of Baking.

*FM Custom Flavors.* Chicago, IL: Food Materials Corporation.

*Foamkill Silicone Antifoam Compounds.* Technical Bulletin. Greenville, SC: Crucible Chemical Co.

*Focus on Food.* Technical Bulletin. Lincolnshire, IL: Purac America.

*Focus General Trends in Fortification.* Technical Bulletin. Lincolnshire, IL: Purac America.

*Food Acidulants.* New York: Pfizer Chemicals Division.

*The Food Chemical New Guide.* Washington, DC: Louis Rothchild, Jr.

*Food Chemicals Codex.* 1972. 2nd ed. Washington, DC: National Academy of Sciences.

*Food Emulsifiers.* Greenwich, CT: Glyco Chemicals, Inc.

*Food Emulsifiers.* Gurnee, IL: Mazer Chemicals, Inc.

*Food Emulsifiers, Coatings, and Lubricants.* 1976. Kingsport, TN: Eastman Chemical Products, Inc.

*Food Emulsifiers and Specialty Products.* 1977. Boonton, NJ: PVO International, Inc.

Food Fortification Aids. 1982. *Food Processing* 43(10): 48.

*Food Grade D-Fructose.* Technical Bulletin. Nutley, NJ: Roche Chemical Division.

*Food Grade Eastman Triacetin.* Technical Bulletin. Kingsport, TN: Eastman Chemical Products, Inc.

*Food Grade Lactic Acid.* St. Louis, MO: Monsanto.

*Food Grade Phosphates, Phosphoric Acid.* Philadelphia: FMC Corporation.

*Food Processing Applications of "Purity Brand" Sodium Caseinate.* Technical Bulletin. Erie, IL: Erie Casein Co., Inc.

*Food Preservatives.* New York: Pfizer Chemicals Division.

*Food Service Seasoning Guide.* New York: American Spice Trade Association.

*Fructose Fruit Sugar.* Helsinki, Finland: Oy Finnsugar Trading Co.

*Fumaric Acid.* 1977. St. Louis, MO: Monsanto.

*Functions of Phosphates in Foods.* Philadelphia: FMC Corporation.

Furia, T. E. 1968. *Handbook of Food Additives.* Cleveland, OH: The Chemical Rubber Co.

*GDL (Glucono Delta Lactone).* Technical Bulletin. Itasca, IL: PMP Fermentation Products, Inc.

*Gellan Gum.* Technical Bulletin. San Diego, CA: The NutraSweet Kelco Company.

*Gluconic Acid.* West Germany: Joh. A. Benckiser GmbH.

*Glucono Delta Lactone.* Technical Bulletin. Itasca, IL: PMP Fermentation Products, Inc.

*Glycine.* Technical Bulletin. Lexington, MS: W. R. Grace & Company.

Goldschmiedt, H. 1974. Why Are Anti-Oxidants Needed in the Food Industry? *Food Trade Review* (March): 10–12.

*GPC Bulletin No. 1004.* Muscatine, IA: Grain Processing Corporation.

Graf, T., and L. Meyer. 1976. Use of Lecithin in Food. *International Flavors and Food Additives* 7: 218–221.

*Grindsted Emulsifiers.* Overland Park, KS: Grindsted Products, Inc.

Griswold, R. M. 1962. *The Experimental Study of Foods.* Boston: Houghton Mifflin Co.

Hannigan, K. J. 1980. Crystalline Fructose: Who's Using It and Why? *Food Engineering* 52(11): 28–30.

Hannigan, K. J. 1980. Spices: Changes Ahead. *Food Engineering* 52(6): 47.

*Hardy High Grade Salt for the Food Industry.* St. Louis, MO: Hardy Salt Co.

*Hawaiian Washed Raw Sugar.* Bulletin. San Francisco, CA: C&H Sugar Company.

Hawley, G. G. 1977. *The Condensed Chemical Dictionary.* 9th ed. New York: Van Nostrand Reinhold Co.

Henderson, J. L. 1971. *The Fluid-Milk Industry.* Westport, CT: AVI Publishing Co., Inc.

Hoch, G. F. 1999. A No-fat Pow-Wow. *Food Processing* 60(7): 60–62.

Hoch, J. G. 1997. Sweet Anticipation. *Food Processing* 58(12): 45–46.

Hoefler, A. C. 1999. Pectins. AACC Short Course. Chicago, IL.

Houston, D. F. *Rice.* St. Paul, MN: American Association of Cereal Chemists.

Hydrogenated Glucose Syrup Permits New Sugar-Free Candies. 1984. *Prepared Foods* 153(3): 158.

*Hysorb®-40.* Technical Data. Denver, CO: Manville Corporation.

*ICP Cocoa Inc.* Camden, NJ: ICP Cocoa Inc.

*The Idea Bank Book.* Kansas City, MO: Patco Products, C.J. Patterson Co.

*Index to Mere-Colors.* Technical Bulletin. North Bergen, NJ: Meer Corporation.

*Information About Silicon Antifoam.* Medland, MI: Dow Corning Corp.

*Ingredients for Food Processing.* St. Louis, MO: Monsanto.

Ingredients Handbook. 1979. *Food Processing* 40(5): 133–151.

*Inulin.* Technical Bulletin. Sugarland, TX: Imperial-Sensus LLC.

*Iron.* 1976. Data Sheet. St. Louis, MO: Mallinckrodt Food Products.

*Iron Sources.* Technical Bulletin. St. Louis, MO: Mallinckrodt Food Products.

*Isosweet® 100 Fructose Corn Syrup.* Technical Bulletin. Decatur, IL: A. E. Staley Mfg. Co.

*Is There Corn in Your Product's Future?* Englewood Cliffs, NJ: Corn Products.

Jones, N. R. 1972. Natural Stabilizers. *Food Technology* 24(12): 626–639.

Jukes, D. 1996. *Food Law.* The Department of Food Science and Technology, The University of Reading, UK.

Kroger, M. 1972, October. Controlling the Quality of Yogurt. Paper presented at Dairy Quality Control Conference.

La Bell, F. 1991. Glycerine Prolongs Shelf Life. *Prepared Foods* 168(8): 69.

Lachance, P. A. 1973. Carbohydrates as Nutrients. *Food Product Development* 7(6): 29–34.

*Lactate Esters.* Technical Bulletin. Holland: C. V. Chemie Combinate Amsterdam C.C.A.

*Lactic Acid and Lactates.* Gorinchem, Holland: C. V. Chemie Combinatie Amsterdam C.C.A.

*Lactic Acid and Lactates.* Technical Bulletin. London, England: Croda Food Products Ltd.

*Lactose.* 1979. San Francisco, CA: Foremost-McKesson, Inc.

*Lactose in Bakery Products.* 1977. Technical Bulletin. San Francisco, CA: Foremost-McKesson, Inc.

Lampert, L. M. 1975. *Modern Dairy Products.* New York: Chemical Publishing Co., Inc.

*Leciflow.* Technical Bulletin. Chicago: Kraft Foods.

*Lecithin.* Technical Bulletin. Chicago: Central Soya Co., Inc.

*Lecithin from Central Soya Naturally.* Fort Wayne, IN: Central Soya Co., Inc.

Lehmann, P. 1979, April. More Than You Ever Thought You Would Know About Food Additives. *FDA Consumer.*

*Lonzest Polyethoxylated Sorbitan Esters.* Technical Bulletin. Fair Lawn, NJ: Lonza, Inc.

Low Calorie Sweeteners. 1983. *Food Engineering* 55(5): 138–139.

*Low Sodium Flavor Enhancers.* Technical Bulletin. Teaneck, NJ: Ajinomoto U.S.A., Inc.

McCormick, R. D. 1975. Aspartame: A New Dimension for Controlling Product Sweetness. *Food Product Development* 9(1): 22, 35–36.

McCormick, R. D. 1983. Food Pigment Use Governed by Broad Considerations. *Processed Prepared Foods* 152(1): 127, 131–132.

McCormick, R. D. 1986. Ginger—The Zing in Asian Cuisine. *Processed Prepared Foods* 155(6): 169–171.

McCormick, R. D. 1982. GRAS Emulsifier Has Broad Spectrum Antimicrobial Properties. *Processed Prepared Foods* 151(8): 125.

McCormick, R. D. 1985. Methods for Developing Unique Textured Foods. *Prepared Foods* 154(3): 155–156.

McDonald, R. W. *Salt.* Chicago: Morton Salt.

*Make Good Foods Better.* Technical Bulletin. Fort Wayne, IN: Central Soya Co., Inc.

*Malic Acid.* Parsippany, NJ: Alberta Gas Chemicals.

*Maltrin®.* Muscatine, IA: Grain Processing Corp.

Marks, J. 1968. *The Vitamins in Health and Disease.* London, England: J&A Churchill Ltd.

Matz, S. A. *Bakery Technology and Engineering.* Westport, CT: The AVI Publishing Co., Inc.

Matz, S. A. 1959. *The Chemistry and Technology of Cereal as Food and Feed.* Westport, CT: The AVI Publishing Co., Inc.

Matz, S. A. *Cookie and Cracker Technology.* Westport, CT: The AVI Publishing Co., Inc.

*Maxarome®.* Des Plaines, IL: GB Fermentation Industries, Inc.

*Meer Natural Colors.* Technical Bulletin. North Bergen, NJ: Meer Corp.

*Minerals for Nutrition.* Technical Bulletin. St. Louis, MO: Mallinckrodt, Inc.

*Monsanto Sorbic Acid and Potassium Sorbate.* St. Louis, MO: Monsanto.

Morris, C. E. 1981. Combines Intense Flavors with Low Salt Content. *Food Engineering* 53(9): 94–95.

Morris, C. E. 1981. FDA Clears Aspartame. *Food Engineering* 53(8): 154–155.

*National Certified FD&C Food Colors.* 1972. Morristown, NJ: Allied Chemical Corporation.

*Natural Fats and Oils.* Technical Bulletin. Boonton, NJ: Drew Chemical Corp.

*Natural Spices.* Madison, WI: Quality Control Spice Co., Inc.

*Natural Vegetable Colors.* Technical Bulletin. Milwaukee, WI: Chr. Hansen's Laboratory, Inc.

Nelson, J. A., and G. M. Trout. 1964. *Judging Dairy Products.* Milwaukee, WI: Olsen Publishing Co.

*Newer Knowledge of Milk.* 3rd ed. Chicago: National Dairy Council.

New Perspective Proven Effective in Tests. 1982. *Food Engineering* 54(11): 60.

New Trade-Offs in Selecting Acidulants. 1974. *Food Engineering* 46(5): 84–85.

*OBI Pectin.* Technical Bulletin. Switzerland: OBI Pectin AG.

Ockerman, H. W. 1978. *Source Book for Food Scientists.* Westport, CT: The AVI Publishing Co., Inc.

*Official Methods of Analysis.* 1975. 12th ed. Washington, DC: Association of Official Analytical Chemists.

Ohr, L. M. 1998. A Sampling of Sweetness. *Prepared Foods* 167(3): 57–63.

*The Oleoresin Handbook.* 1974. New York: Fritzsche Dodge & Olcutt, Inc.

*On Lactose.* San Francisco, CA: Foremost-McKesson, Inc.

Paul, P. C., and H. H. Palmer. *Food Therapy and Applications.* New York: John Wiley & Sons, Inc.

*Peanut Oil.* Tifton, GA: Georgia Agricultural Commodity Commission for Peanuts.

Peckham, G. C. 1964. *Foundations of Food Preparation: A Beginning College Text.* New York: The Macmillan Company.

*Penick Plus-A-Tives® and Flavor Materials.* New York: S. B. Penick & Co.

Petrowski, G. E. 1976. Emulsion Stability and Its Relation to Foods. Reprinted from *Advances in Food Research* 22.

Petrowski, G. E. 1975. Food-Grade Emulsifiers. *Food Technology* 29(7): 52–62.

*Phosphates.* 1979. Technical Bulletin. Philadelphia: FMC Corporation.

*Phosphoric Acid.* Westport, CT: Stauffer Chemical Co.

*Polydextrose.* 1982. New York: Pfizer, Inc.

*Pomalus.* 1977. Morristown, NJ: Allied Chemical Corp.

*Potassium Chloride.* Westwood, NJ: Tri-K Industries, Inc.

*Potassium Gluconate.* Technical Bulletin. Itasca, IL: PMP Fermentation Products, Inc.

*Potassium Hydrogen Tartrate.* FR Germany: Joh. A. Benckiser GmbH.

*Products Catalog 1977–1978.* Harbor City, CA: Hathaway Allied Products, Inc.

*Product Guide Cocoa Powder.* 1974. Holland: Cacaofabriek de Zaan.

*Product Information on Phosphates.* Technical Bulletin. Westport, CT: Stauffer Chemical Company.

*Promine®-F.* Chicago: Central Soya Co., Inc.

*Promine®-R.* Chicago: Central Soya Co., Inc.

*Promosoy®-100.* Chicago: Central Soya Co., Inc.

*Properties of Kelcogel® LT100 Gellan Gum.* Technical Bulletin. San Diego, CA: The NutraSweet Kelco Co.

*Propylene Glycol.* New York: Union Carbide Corp.

*The Protein Power of Soybeans.* Chicago: Food Protein Council.

*Proteins from EEC.* Erie, IL: The Erie Casein Company, Inc.

Przybyla, A. 1980. Colors. *Processed Prepared Food* 149(9): 109–114.

Przybyla, A. 1983. Oils: Oil Selection Key to Many New Products. *Prepared Foods* 152(4): 97–100.

Przybyla, A. 1981. Polydextrose New Food Additive Approved by FDA. *Processed Prepared Foods* 150(8): 75–76.

Pszczola, D. E. 1997. Curdlan Differs from Other Gelling Agents. *Food Technology* 51(4): 30.

Pszczola, D. E. 1999. Sweet Beginnings to a New Year. *Food Technology* 53(1): 70–76.

*Pyridoxine Hydrochloride.* Technical Bulletin. New York: Fallek Sales Inc.

*Questions and Answers About Phosphates in Food.* Westport, CT: Stauffer Chemical Company.

Refined Fish Oils Hold Potential for Food Fortification. 1998. *Food Technology* 52(4): 76.

*The Remarkable Story of Monosodium Glutamate.* Washington, DC: International Glutamate Technical Committee.

Riboh, M. 1977. Natural Colors: What Works . . . What Doesn't. *Food Engineering* 49(5): 67–72.

*Ribotide®.* Tokyo: Takeda Chemical Industries, Ltd.

Rice, J. 1983. Yucca Plant Extract. *Food Processing* 44(4): 58.

Robinson, R. F. 1972. What is the Future of Textured Protein Products? *Food Technology* 26(5): 59–63.

*Roche Carotenoids*. 1986. Technical Bulletin. Nutley, NJ: Roche Chemical Division.

Rust, R. E., and D. G. Olson. *Meat Curing Principles and Modern Practice*. Kansas City, MO: Koch Supplies, Inc.

*SAG® Silicone Antifoams*. 1975. New York: Union Carbide.

Sales Soar for Carob Powder Used in Candies, Cookies, Cakes, Instant Drinks . . . 1977. *Food Engineering* 49(9): 43.

*Salt in the Meat Industry*. St. Clair, MI: Diamond Crystal Salt Co.

Sanderson, G. R. 1994. *Gellan Gum*. Technical Bulletin. San Diego, CA: Kelco Division of Merck and Co.

Saussele, H. Jr. 1982. Trends and Future of Yeast. *Food Engineering* 54(5): 129.

*Sequestrene Na$_2$ Food Guide and Na$_2$Ca Food Grade EDTA*. 1985. Greensboro, NC: Ciba-Geigy Corporation.

Signorino, C. A., and E. R. Furmanski. 1975. Dyes Cause Color Problems? Try Certified Lakes. *Food Engineering* 47(5): 76–77.

*Silicone Antifoams*. 1973. Skokie, IL: Hodag Chemical Corp.

Sinclair, P., R. S. Vettel, and C. A. Davis. *Soybeans in Family Meals*. Washington, DC: U.S. Dept. of Agriculture.

*Sodium Bicarbonate*. 1987. New York: Church & Dwight Co., Inc.

*Sodium Bicarbonate*. 1982. Technical Bulletin. Morristown, NJ: Allied Chemical.

*Sodium Citrate in Foods*. 1969. New York: Pfizer Chemicals Division.

*Sodium Saccharin Powder*. 1981. Chicago: Pettibone-Chicago, Inc.

*Solka-Floc®*. New York: Brown Company.

*Sorbic Acid and Potassium Sorbate*. St. Louis, MO: Monsanto.

*Sorbitol*. 1972. Technical Bulletin. Fairlawn, NJ: Lonza, Inc.

*Sorbitol in Foods*. 1975. New York: Pfizer Chemicals Division.

*Sorbitol, Sorbitol*. 1971. New York: Pfizer Chemicals Division.

*Speciality Grains Add Interest and Nutrition*. Portfolio. Fresno, CA: Sunnyland Mills.

*Specification for the Identity and Purity of Food Additives and Their Toxicological Evaluation: Emulsifiers, Stabilizers, Bleaching and Maturing Agents*. 1964. Geneva, Switzerland: World Health Organization.

*Spices, Flavors, Flavor Characterization*. 1966. New York: Fritzche Dodge & Olcott, Inc.

*SPI Polyols*. Technical Bulletin. New Castle, DE: SPI Polyols, Inc.

*Stadex© Dextrins*. Technical Bulletin. Decatur, IL: A. E. Staley Manufacturing Company.

*Starch.* Technical Bulletin. Bridgewater, NJ: National Starch and Chemical Company.

*Starch.* Technical Data. Decatur, IL: A. E. Staley Mfg. Co.

*Star-Dri® Corn Syrup Solids.* Decatur, IL: A. E. Staley Mfg. Co.

*Sta-Sol® Lecithin Concentrates.* Technical Bulletin. Decatur, IL: A. E. Staley Mfg. Co.

*Sta-Sol® R Lecithin Compound.* Technical Bulletin. Decatur, IL: A. E. Staley Mfg. Co.

Stephenson, M. G. 1975. Textured Plant Protein Prodcuts: New Choices for Consumers. *FDA Consumer* 9(3): 18–23.

*Straight Talk About Salt.* 1982. Alexandria, VA: Salt Institute.

*Strike Oil.* Bismarck, ND: North Dakota Sunflower Council.

Sugarless Syrup. 1984. *Food Engineering* 56(5): 68–69.

*Sunflower.* 1982. Bismarck, ND: National Sunflower Association.

*Sustane® Food-Grade Antioxidants.* Technical Bulletin. Des Plaines, IL: UOP Process Division.

*Syloid®.* Baltimore, MD: W. R. Grace & Co.

*Syloid® Conditioning Agents for the Food Industry.* Baltimore, MD: W. R. Grace & Co.

*Syncal® Calcium Saccharin.* 1981. Cleveland, OH: Sherwin-Williams Company.

*Syncal® Sodium Saccharin.* 1981. Cleveland, OH: Sherwin-Williams Company.

*Takeda-Fallek Sales, Inc.* Technical Bulletin. New York: Takeda-Fallek Sales, Inc.

*Tannic Acid.* Technical Bulletin. Flushing, NY: Aceto Chemical Co., Inc.

*Tenox® Food-Grade Antioxidants.* 1974. Kingsport, TN: Eastman Chemical Products.

*Terra Alba, Calcium Sulfate Dihydrate.* Technical Bulletin. Elkhart, IN: Miles Laboratories, Inc.

Testing Time for Tofu. 1984. *Food Manufacture* 59(5): 76–77.

Triebold, H. O., and L. W. Aurand. 1969. *Food Composition and Analysis.* New York: Van Nostrand Reinhold Co.

*Tripotassium Citrate.* FR Germany: Joh. A. Benckiser GmbH.

*Trisodium Citrate.* FR Germany: Joh. A. Benckiser GmbH.

Tungland, B. C. 1998. A Natural Pre-biotic. *The World of Ingredients* (September): 38–41.

Twigg, B. A. 1974. *Ingredient Technology for Product Development.* Portland, OR: IFT Short Course.

Valley Evaporating Company. Specification Sheet #21. Yakima, WA: Kingblossom Apple Flakes & Powders.

*Veltol®-Plus and Veltol®.* New York: Pfizer Chemicals Division.

*Versene® Ca and Versene® NA Food Grade Chelating Agents.* Technical Bulletin. Midland, MI: Dow Chemical Co.

*We Make Things Taste Good.* Technical Bulletin. Camden, NJ: Mafco Worldwide Corp.

*Wheat Gluten.* 1981. Shawnee Mission, KS: International Wheat Gluten Association.

*Wheat Starch.* Technical Bulletin. Minneapolis, MN: Henkel Corp.

Whistler, R. L., and J. N. Bemiller. 1973. *Industrial Gums.* New York: Academic Press.

Windholz, M. 1976. *The Merck Index.* 9th ed. Rahway, NJ: Merck & Co., Inc.

Woodin, G. B. 1981. The Elegant Egg. *Food Product Development* 15(4): 44–56.

*Worcestershire Sauce.* Technical Bulletin. Dallas, TX: The Illes Co.

*The World of Norda Flavors.* New York: Norda Essential Oil & Chemical Co., Inc.

Wouters, R. 1998. The Benefits of Inulin and Oligofructose in Ice Cream. *The World of Ingredients* (September): 44–45.

Xylitol: Sugar's New Rival. 1976. *Business Week,* September 3, p. 765.

*Yeasts for Wine Making.* 1981. Bulletin. Milwaukee, WI: Universal Foods Corporation.

*Your Guide for Buying Malt Extracts and Related Syrups.* Milwaukee, WI: Premier Malt Products, Inc.

Ziemba, J. V. 1975. Food Protein Research Accelerates. *Food Processing* 36(8): 21–27.

Zind, T. 1999. Faking the Fat. *Food Processing* 60(8): 54–59.

Zind, T. 1999. Stay Keen on the Sweetener Scene. *Food Processing* 60(7): 54–58.

Zoller, J. M., et al. 1980. Fortification of Non-Staple Food Items with Iron. *Food Technology* 34(1): 38–46.